METHODS ON PHYSICO-CHEMICAL ANALYSIS OF FRUITS

Methods on Physico-Chemical Analysis of Fruits

Dr. B.C. Mazumdar

Dr. K. Majumder

2021

Daya Publishing House®

A Division of

Astral International Pvt. Ltd.

New Delhi - 110 002

First Published, 2003
Reprinted, 2021

ISBN: 978-81-7035-288-4 (Hardbound)

Publisher's note:

Every possible effort has been made to ensure that the information contained in this book is accurate at the time of going to press, and the publisher and author cannot accept responsibility for any errors or omissions, however caused. No responsibility for loss or damage occasioned to any person acting, or refraining from action, as a result of the material in this publication can be accepted by the editor, the publisher or the author. The Publisher is not associated with any product or vendor mentioned in the book. The contents of this work are intended to further general scientific research, understanding and discussion only. Readers should consult with a specialist where appropriate.

Every effort has been made to trace the owners of copyright material used in this book, if any. The author and the publisher will be grateful for any omission brought to their notice for acknowledgement in the future editions of the book.

Published by : **Daya Publishing House®**
A Division of
Astral International Pvt. Ltd.
– ISO 9001:2015 Certified Company –
4736/23, Ansari Road, Darya Ganj
New Delhi-110 002
Ph. 011-43549197, 23278134
E-mail: info@astralint.com
Website: www.astralint.com

Printed at : **Replika Press Pvt. Ltd.**

Dedicated to the venerated memory
of the
Indian Scientist and Philosopher,

ACHARYA
JAGADISH CHANDRA BOSE
(1858-1937)

Preface

Fruits that are consumable to the human beings are gifted with a number of attributes of qualitative and nutritive importance. The attributes are not the same for all fruits and vary in accordance with species, horticultural varieties, eco-environmental conditions of growing the trees bearing them and the inputs used in their culture. A fruit scientist aims at studying these principles in fruits as influenced by the above factors either in the pre-harvest stage or under post-harvest condition and hence, the methods involved in qualitative or quantitative determination of them is evidently, a pre-requisite to conduct such studies.

Although there exists a number of books which have presented the methods of analyses of agricultural and horticultural produce, any separate title dealing exclusively on the analytical procedure of fruits under Indian conditions is lacking. But it needs pointing out in this context that in properly conducted analytical studies in fruits, it becomes imperative to test a number of characteristic features which claim speciality in themselves unlike materials of other kind. Further, these characteristically identified compounds are not the same for all fruits and may have specificity in accordance with the factors mentioned above. Analyses of the components from chemical standpoint though authenticate these distinguishing features, for some of the facets, even senses of observations, feeling, attitude of human beings also serve as valuable measuring indices and hence, cannot be set at defiance.

In consideration to the above understanding, the present title has been attempted to provide the procedural aspect in regard different traits which make their presence in the human consumable fruits. However, it needs pointing out that the ingredients in different fruits if are enlisted, the number will have no bound. Among such multitudinous physical and chemical components of fruits, only a few have been sorted out in the book which are more commonly considered in fruit laboratories for their greater applicability in assessing quality, nutritive value and acceptance to consumers. Practicability and less sophisticated methodology have also been duly considered in stating the procedural details.

The title has primarily aimed at meeting the requirements of the students, analysts and researchers in the areas on horticulture, promology, food technology, human nutrition, plant biochemistry, botany, applied botany, forestry, pharmaceutics, ayurved and other disciplines who are concerned with analyses of fruits. Besides, need of the horticultural and fruit processing laboratories, factories and training centres in the private or public sector and that of the agricultural marketing personnels in the act of grading of fruits has also been kept in mind by the authors. However, it needs pointing out that many of the methods presented in the title may be applied for constitutional analyses in plant organs other than fruits also with necessary modifications.

Authors will consider their work successful if this practical manual is able to render any sort of help to its user and like any author of a scientific title, they always welcome valuable suggestions from anybody for making improvement of the future editions of the title.

Above all, the authors would consider failing their duty, if they do not gratefully acknowledge the books and other published reports presented by different authors, for without consultation of those, the present title would not have come to a reality.

November 30, 2001   **B. C. Mazumdar**
Calcutta University   **K. Majumder**

Contents

General Precautions to Work in the Laboratory and Field

As safeguard measures against any accident, the fruit analyst should be extremely careful at every step in his work whether in the laboratory or in the field. Some of the common precautions which he should especially keep in mind have been stated below.

In the Laboratory :

Alertness

The most important point for the laboratory worker is to remember the fact that he should always be on the alert while entering into the laboratory just as a motor car driver when he holds the steering wheel to start driving and a surgeon when he is ready to open the abdomen of his patient. The laboratory worker should keep in mind the 4-A slogan, *i.e.*, Always Alert, Accident Avert.

Provision of emergency equipments

Before starting the work, the analyst should check and get confirmed that the emergency equipments, e.g., the fire-extinguishing set, sand and water-filled buckets, first-aid box etc. are in proper place and that those are in working condition. Installation of telephone set in the laboratory is definitely of advantage and some telephone numbers, especially those of fire-stations, first-aid personnel, physicians, ambulance centres, hospitals as well as electricians, gas-operators, fitters, plumbers etc. should be legibly displayed on boards and hung in wall at such place as may be easily noticeable to anybody.

Avoidance to work alone

At least another person should remain present in the laboratory as long as the worker continues to work there. If not, the person should be present close to the laboratory which the worker should know and from where he may rush to the laboratory in emergent situation.

Proper dress

The worker should never come to the laboratory with garments made up of polyester, silk or such materials which are highly inflammable. Apron should be a must to the worker and a cap on the head is also advisable, especially for the ladies maintaining long hair. If the spectacles is not used by the worker, he should keep with him one set of powerless glasses in order to protect his eyes. It should be noted in particular that the ultra-violet ray with which he may have to work for disinfection of site or for observing chromatograms is liable to cause permanent damage to the eyes on receiving direct exposure. Use of a spectacles may however, act as a shield for glass being impermeable to this radiation.

Ventilation

Especially in a chemical laboratory, the air contains pollutants which sometimes reach so high degree that may seriously act upon the health of the worker, especially in his cardio-vascular or respiratory system. Before entering into the laboratory, it is therefore, advisable that the exhaust fans, ceiling fans, air-circulators, pollutant absorbing instruments etc. should be switched on and all the windows be opened. These should be left in this way for some time to drive away the harmful gases and then the worker should enter there. At times, enriching the oxygen level of the laboratory atmosphere by pumping in of the gas or by electrical device is also a healthful practice, if possible to be done.

Supply lines

Before starting the work, it is also an essential step to check the electric, gas and water connections to ensure that their supply continues, have no leakage problem, bare wire etc. and can be operated safely.

Storing of chemicals

Concentrated acids, alkalis and other corrosive chemicals should be stored separately at a safe place in tightly stoppered bottles or jars and not at a high level from the floor. Organic solvents like ether, chloroform, acetone, alcohol, acetic and other organic acids etc., which evaporate quickly, should be kept in extra-tightly stoppered bottles. These organic chemicals being highly inflammable should always be stored sufficiently away from fire source. The worker should note that no single chemical or even distilled water should be stored in the laboratory unless it is properly and legibly labelled.

Handling of chemicals

Any chemical should not come in direct contact with the body of the worker. If it so happens accidentally, the body part should immediately be thoroughly washed by irrigating with running tap water.

Heating of organic solvents or other inflammable regents should be done only over water- or sand-bath as the case may be and maintaining temperature lower than that at which they catch fire. Digestion in acids should also be done in specific chamber meant for the purpose.

The dangerous practice of pipetting by mouth concentrated acids, alkalis, toxic solvents or solutions or any liquid which gives off harmful fume should never be done.

For the purpose of diluting concentrated inorganic acids with water, the acid should be poured slowly and gradually in water slantwise, instead of following the reverse process.

Liquor ammonia bottle should not be opened without getting chilled. It should be opened only in suitable chamber taking extreme care that the content does not gush out in spurt and come to the eyes of anybody which may be permanently damaged by this.

Solutions of hydroxide pellets in water which result exothermic reaction should be done by stirring very slowly and carefully. Instead of preparing such a solution every time it is needed, it is advisable to prepare a concentrated stock solution and store it in tightly stoppered bottle. When required, this may be used by making desired dilutions by titrating against a standard acid.

As emergency measure, it is advisable to maintain some dilute acids and alkalis at a convenient place in the laboratory .These may be used to neutralize any acid or alkali, if accidentally falls on the working table or the floor.

Handling of glass or plastic materials

Any glass-or plastic-ware which is broken or crack should never be used. For keeping concentrated acids, alkalis etc., containers made up of low-graded glass that is highly fragile should be avoided. Heating of any solution should not also be done in container of such type of glass and if plastic-ware is used, it should not be heated without knowing the heat withstanding capacity of it.

To work with glass-wool, utmost care should be taken that it never comes in contact with eyes or any part of the worker's body. It should be noted that even highly sophisticated operative procedure may also fail to pick out all the particles from the body part, if these are pierced there.

Handling of instruments

A laboratory instrument should not be handled unskillfully and without knowing the *modus operandii* of it. If any heating equipment is used, the temperature should regularly be recorded so that the desired limit is not exceeded. The door of a muffle- furnace should not be opened when it has attained high temperature for such an act may burn the face or hand of the worker. It should be opened and ashed materials be removed from it only when the temperature has come down to that of the room.

Cleanliness in work

Cleanliness is said to be next to godliness. In a scientific work, cleanliness not only paves the way for methodological correctness so as to bring about accuracy of results but is also a factor that is very much responsible to avert accident, which the worker should not forget.

Disposal of broken glass-pieces

If any glass-ware is broken, the site should immediately be swept and cleaned leaving no trace of particle there. Dilly-dallying in this matter may lead to serious injury to the worker or others.

Use of radio-active chemicals

To work with any radio-active chemical, special training is essential for the worker. He should take utmost precautions at every step to protect himself and of others. The disposable materials contaminated with radio-chemical should be carefully collected instead of throwing here and there, and disposed as directed by the appropriate authority. It should be remembered that contaminating others with radiations amounts to a legal offence. The worker himself should also record regularly the degree of radiation received by him to ensure that it has not exceeded the permissible limit.

Avoidance to eat or drink during work

Eating or drinking while doing laboratory work may lead to a serious disaster and hence, the worker should refrain from taking even a glass of water without washing his hands properly.

Actions in case of accident

Precautions avert accident but even with utmost precaution, if any accident occurs, it should be tackled with effective steps depending on the type. The first point that is to be remembered by the worker is that he should be guided by his own *cc 'nmon sense,* ready wit, judgement and patience when an accident has taken place. It should not be forgotten that a small fire caught by the clothes of a person may burn him completely and result in a conflagration, if he runs helter-skelter while he may be saved, if he is guided by his common sense, which would advise him to strip off the garments as quickly as possible without any idiotic shyness. Similarly, irrigating the body part with running tap water should be the first step in case of contact with any acid or alkali. However, it is an admitted fact that an accident victim very often gets nervous and is at what to do. Hence, he should quickly inform the person who is present in the laboratory or close to it about the happening. It will be in fact, much easier by others to take effective steps.

Cleaning before leaving

The worker should not forget the fact that either himself or some other person may come to his table to work when he leaves the place. Accordingly, it is his bounden duty as a scientist to clean up the working table, desk etc. and ensure that no chemical or solution remains fallen here and there. The used glass- or plastic-wares need also be washed thoroughly just after work, instead of storing them. Before leaving the laboratory, he should also make sure of the fact that the electric switches and gas-burners are put off, water-taps are turned off and the instruments used have been properly kept.

In the Field :

While working in the field or orchard also, the fruit analyst should be careful against any possible accident. Some of the common precautionary measures that are to be taken to work there have been stated below.

The worker should visit the area carrying with him always some materials which may be of great help to him at any time. These are, a strongly built stick, a good knife, a bundle of thread, some absorbent cotton, bandage, sticking plaster, antiseptic lotion or cream etc. Wherever there is danger of venomous snakes, it is also advisable to carry anti-venom serum in a thermos containing ice along with sterilized syringe and some rectified spirit.

It is always advisable for him to put on a long pair of strongly built shoes and preferably, a pair of jack-boot, which cover legs upto the knees. This is extremely useful against attack by snakes, scorpions or other venomous ground animals. Throwing a little amount of diluted commercial phenol at the site some time before visiting there is also an effective means to repel snakes and other animals. The worker should also strike the stick on the ground while he moves for vibration of this sound may dispel snakes and other animals. Application of mosquito and insect repelling lotion or oil on hands and feet is also helpful where necessary.

Reaching greater height of a fruit tree should never be attempted without using a strongly built, self -supported ladder. It is still better if the ladder is provided with railings at the top and along the rungs of it, in order that the worker does not sustain any fall from the ladder.

If the worker does not use spectacles, he must use one set of powerless glasses. This is especially helpful to work at upper height of a tree. As in such case he has to look above every now and then, the spectacles would protect his eyes against fall of any dust particle, insect, flower anther etc. The necessity of the spectacles is felt much when an agro-chemical becomes necessary to be applied at upper height of the tree.

A hooded hat is no doubt of much advantage to protect the scalp of the worker not only against scorching heat of sun but also against falling of any biotic or neuter material over it.

If the worker sustains a cut on any part of his body, he should immediately apply antiseptic drug on it and cover the part thoroughly. He should note that procrastination may lead to a serious turn even if the cut is a very minor one. It should also be remembered by him that working in the field without being inoculated by tetanus vaccine with recommended dose may be regarded to be a life-risk factor .

The worker should ensure that somebody else remains present at the site as long as he works there. In any case, the person should be present within easy reach of the worker such that he may be of immediate help to the worker in case of any emergency.

1

Collection of Fruit Samples

1.1 Selection of Fruits

Selection of fruits which are to be plucked from a fruit tree for the purpose of conducting physical or chemical study of them is an operation that needs critical attention and should never be given less importance. This is because, any error committed in this step may be responsible for incorrectness of the ultimate results.

It is a general observation that branches of the common fruit trees or shrubs extend over a relatively large area depending on the species and varieties and that they bear fruits both in the interior and the exterior, *i.e.*, the peripheral parts. In many trees or shrubs, the fruits are grown on all their sides while in others, they are not found to be produced at one or more sides. Again, failure to bear uniformly on all sides is also sometimes noticed.

Beyond doubt, it is a ridiculous idea if anyone goes for plucking all the fruits borne by a tree with the idea to make a test of them, particularly when a few fruits are sufficient for this purpose. Therefore, the practical approach is to pluck only some fruits from the tree and subject them to study.

Now the question is, which fruits are to be plucked for the purpose, *i.e.*, for carrying out the physical or chemical examination. As a matter of fact, if any specification is given, plucking of the fruits should be done keeping in view of that. For example, say, if a fruit scientist has the objective to make a comparative study between the fruits that are borne at the northern and southern sides of a tree, he should take fruits from these two sides only and maintain these two lots separately for carrying out the test. Similarly, if the objective is to make a comparison between the internally and externally borne fruits, he should collect fruits which are produced in the inside and the peripheral parts respectively and in this case also, the fruits plucked from these two parts should be maintained separately to conduct study.

On the other hand, if no such specification has been given, it leaves a question to the fruit analyst as to which fruits are to be considered for plucking. In other words, if the objective of the study is only physical or chemical assay of the fruits borne by a tree and has no other specification, which fruits should actually be plucked for conducting is necessary to be known.

There are in fact, two viewpoints among the fruit scientists on this aspect. One group of scientists is of the opinion that since only a few fruits are to be plucked from the tree which has actually produced a large number and evidently, when the few fruits that would be plucked would *represent* all fruits of the tree, fruits should be plucked from all parts of it. In other words, fruits which though may be few in number need to be procured from as many parts of it as possible, such as from the lower and the upper parts, interior and exterior parts and including those that are borne regardless of sides, *e.g.*, north, south, east, west, etc. According to them, a representation of all fruits produced by the tree can be done when the fruits are taken from all sides.

Another group of fruit scientists hold the view that since the fruits which have been borne in the interior part of a tree usually do not receive adequate amount of sunshine and air and since only those fruits which are borne at the peripheral sides of the tree receive that, the latter would only be procured for carrying out any study. According to this group, the fruits which have been produced on all the peripheral sides should only be used for carrying out any physical or chemical test of them. As a matter of fact, the analyst should mention specifically from which of the tree(s) he has actually collected the fruits to make the representative sample for conducting the study.

Randomization in Selection

A vital point to remember in connection with collection of fruits for any analytical study is that, fruits which are to be plucked should be done only at *random basis,* in other words, without making any biasness in their selection. Thus, if fruits are to be procured say, from the northern side of the tree, those should be plucked randomly from that side. Likewise, if peripherally borne fruits are desired to be obtained, those which have been produced on all the peripheral sides should be plucked at random basis without making any preferential selection.

1.2 Method of Plucking

How to pluck the fruit for the purpose of conducting the test with them is a step that should be done with great care. In fact, two aspects should be given emphasis in this process. These are, firstly, the fruits should not be broken, damaged or bruised in plucking and secondly, a fruit which is to be plucked should be done together with some part of foliage and in any case, the fruit bearing stalk or the pedicel where present must remain attached to it. This is because, any damage done to a fruit may be a source of infection to it and secondly, a fruit which is stripped of its stalk is usually more prone to be ripened quickly.

Plucking of a fruit should actually be done with a sharp cutting instrument which gives a clean and smooth cut on the part and hence, cutting with a knife or pulling a fruit by hand is not considered to be a correct procedure.

For plucking of such fruits which are within the reach of the hands of the harvester, a *secateur* is undoubtedly an ideal instrument in most cases. Various types of secateurs (Figure 1) are available in the market and the heavier type is able to cut a green-wood branch of upto 1 cm diameter.

However, if the fruits that are to be obtained are actually located at much greater height, the secateur is difficult to use unless the harvester climbs on a self-supported lifting device, but there may be risk to the harvester in this as well. In such case, as shown in Figure 2, a fruit-plucking implement, which is also used for truncating and pruning of upper branches may be used with convenience. The implement can be fitted with a pole and thus, the harvester can easily pluck the fruits standing on the ground level holding the pole in suitable position and pulling the rope, the other end of which has been tied to the movable knife of it. The fruits along with shoots which are to be cut will get collected in the cloth-bag which is suitably attached to the implement.

1.3 Sorting

Since the fruits for analyses are collected at random basis, not only the healthy ones but those which are unhealthy, *i.e.*, those

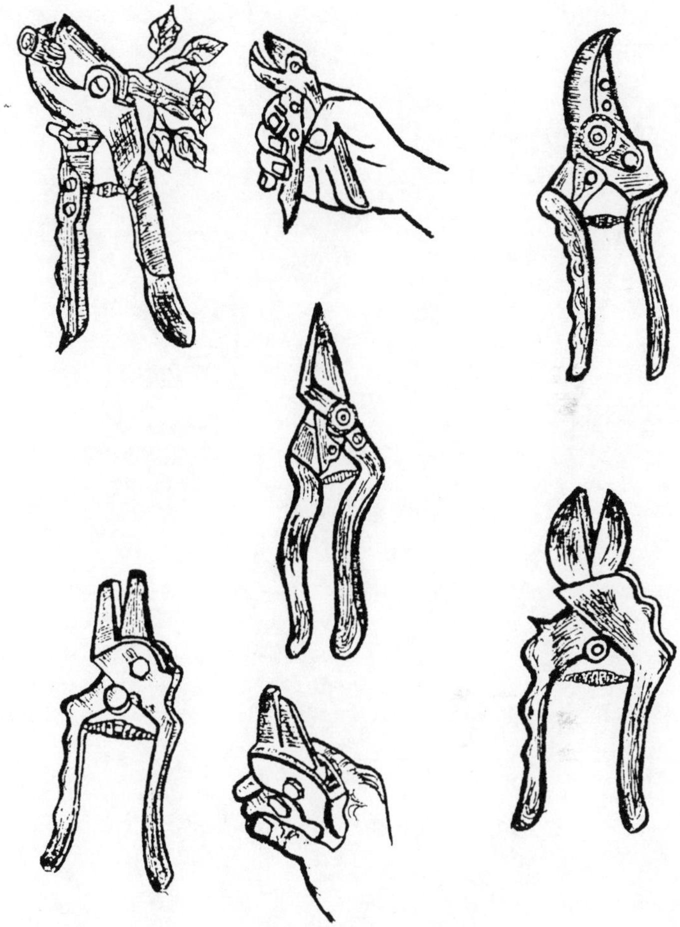

Fig. 1: Different types of secateurs.

infected with insect pests or diseases, or mechanically damaged may also be plucked. Unless otherwise specified, those of the latter group are to be discarded as the same are not considered suitable for experimental use.

The infection occurring in the inside of a fruit may however, be difficult to be detected without cutting it but that on its outer surface can be known easily on visual observation of symptoms. At the first step therefore, the fruits showing any symptom of infection or damage on surface should be scrupulously rejected such that

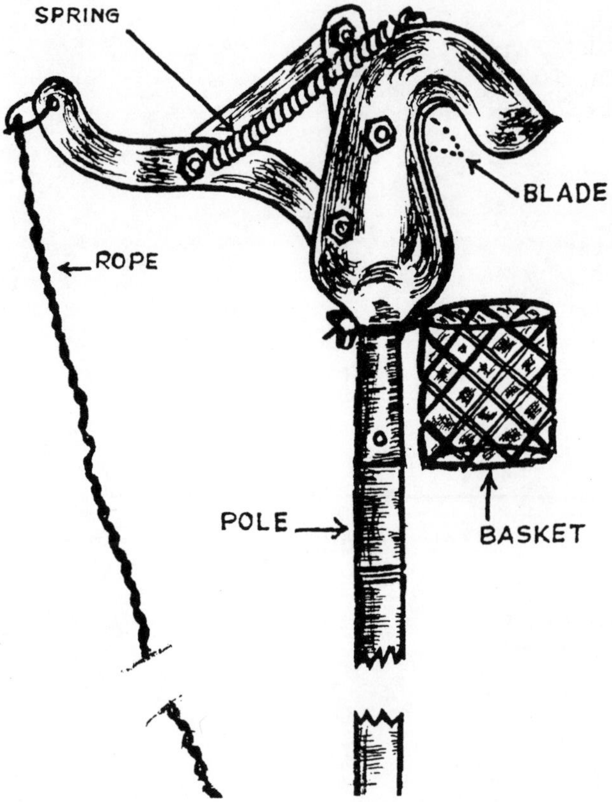

Fig. 2: Fruit plucking implement.

only those which at least outwardly look healthy are only sorted out for conducting the study.

1.4 Surface Cleaning

Especially when fruit trees are grown outdoor in open situation, not only their leaves, branches and flowers but the fruits borne by them also being exposed receive dirt, dust, excreta of insects, squirrels etc. on their surfaces. Evidently, such fruits cannot be used for analytical purpose without cleaning them to make free from those foreign materials adhered on their surfaces. Therefore, it is a necessary step that after cutting off the stalks or other foliage parts which remain attached to the fruits after harvest, they should be meticulously cleaned before use.

Surface cleaning of a fruit can be done by rubbing it with a piece of dry clean cloth, or in a better way by rubbing with moistened and then with a dry cloth-piece. However, if this procedure is not enough to clean it properly, washing with clean and preferably running water may be resorted to (Figure 3). Washing a fruit for unnecessary long period should in any case be avoided as it may cause to absorb some amount of water, whatever small may be. By this, there may be a change of its original weight and concentration of soluble components. After washing, it is a necessary step to wipe out the water adhered on the surface of a fruit. This may be done by rubbing it lightly with a piece of dry, clean, and soft cloth, flannel, blotting paper, sponge, absorbent cotton etc. before subjecting it to assay.

Without wetting a fruit, the dirt and dust or other loosely adhered foreign materials can also be removed from a fruit surface with the help of a suitable suction apparatus, *e.g.*, a vacuum-cleaner but tightly stuck materials are difficult to be effaced by this.

Surface cleaning of fruits may be done right in the orchard itself after plucking them, or after those are brought to laboratory for analysis.

1.5 Bringing to Analytical Laboratory

In many cases, the laboratory where the fruits are to be analyzed is somewhat away from the site where the trees bearing them are actually growing. There are instances in some parts of India where the analytical laboratory is at long distance from the experimental plantation of fruits. Therefore, the fruits that have been collected need to be carried to the laboratory for conducting the study and this will involve some time.

During the time that is taken in transit, there is however, likelihood that the fruits lose weight, shrink or undergo metabolic changes in them. Under post-harvest condition, these changes usually occur at a faster rate and more so with increase of warmth. Therefore, if the fruit analyst desires to examine the physical or chemical condition of the fruits as they would have been at the time of harvest, he does not get correct experimental materials to conduct the analysis.

To avoid this, attention should be paid on two aspects in particular. These are firstly, analysis of the fruits should be done as

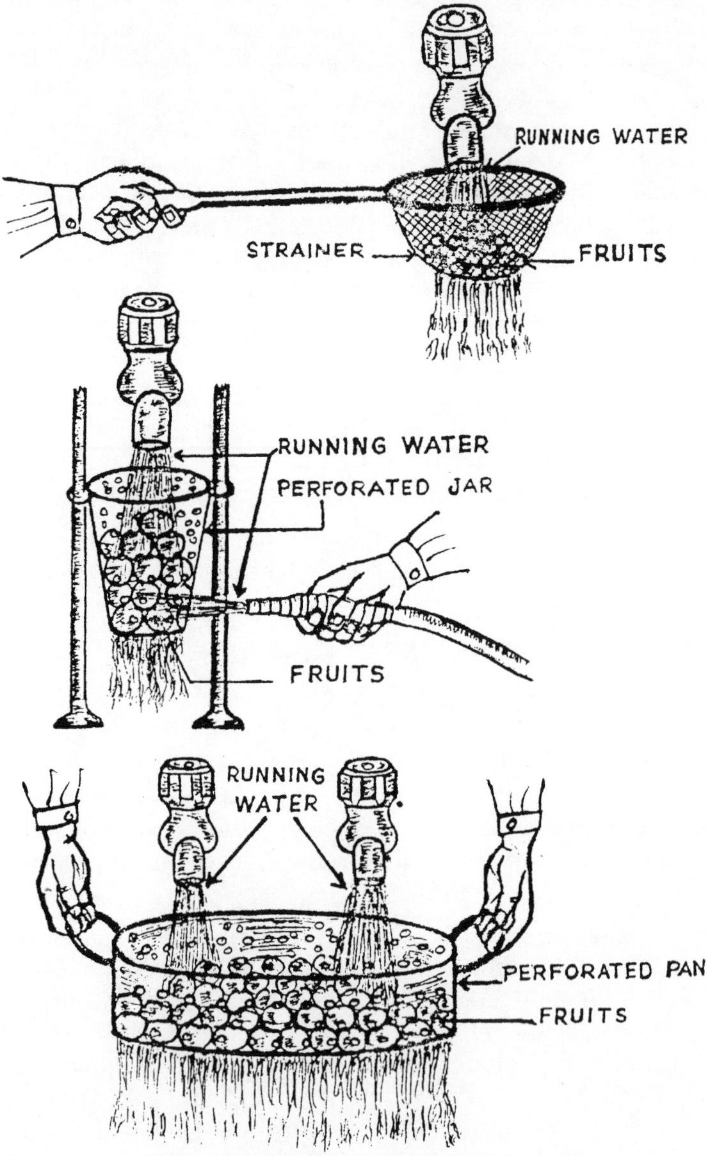

Fig. 3: Washing fruits in running water.

soon as possible after they are plucked and secondly, the atmosphere to which the fruits are exposed in transit or subsequent storage should be so maintained as would result in minimal change of the physio-chemical condition of them. Among other factors, lower temperature with or without higher relative humidity of the atmosphere depending on fruits belonged to different species and varieties are known to be congenial condition for this.

In accordance with, carrying the fruits in a refrigerated box, ice-box, etc. appear to be most suitable. In unavailability, a suitable thermos box or flask may be used and in absence of them, a box made up of insulating material should at least be used. Covering the fruits with two polythene packets of thicker gauge, interposed with moist cloth or cotton may even be better than carrying the fruits in an ordinary bag to the laboratory.

2

Making Representative Sample of Intact Fruits

In order to obtain scientifically acceptable results, conducting any analysis using a single fruit is not justifiable, unless however, such an objective of the study, has been specified. As a matter of fact, a number of fruits that belonged to the identical category should be considered together as a whole to consider as one unit to perform the study. Thus, an analytical sample is required to be made for this purpose and the sample should in fact, comprise a minimum number of fruits.

Accordingly, if fruits of a given tree are required to be tested, few fruits that have been borne by that tree should be used and not one fruit only. Now, which fruits are to be actually used is again a question to the analyst. The answer is, the analyst should be guided by the principle that the sample which he would make with several fruits should actually *represent* the lot, considered for the study. Thus, among many fruits that have been produced by the tree, the few which have been taken for the study, should represent all fruits of the tree as far as possible. The sample may therefore, be termed as the *representative sample*. Similarly, say if fruits borne at the northern side of a tree are needed to be tested, the representative sample should be made with several fruits which are borne at the northern side.

Procedure

In making a representative sample of fruits, *randomization* is the most important aspect that should be kept in mind and followed at every step. Among other procedures that are followed, one which is given below appears to be simple, practicable in most cases and does not require any instrument.

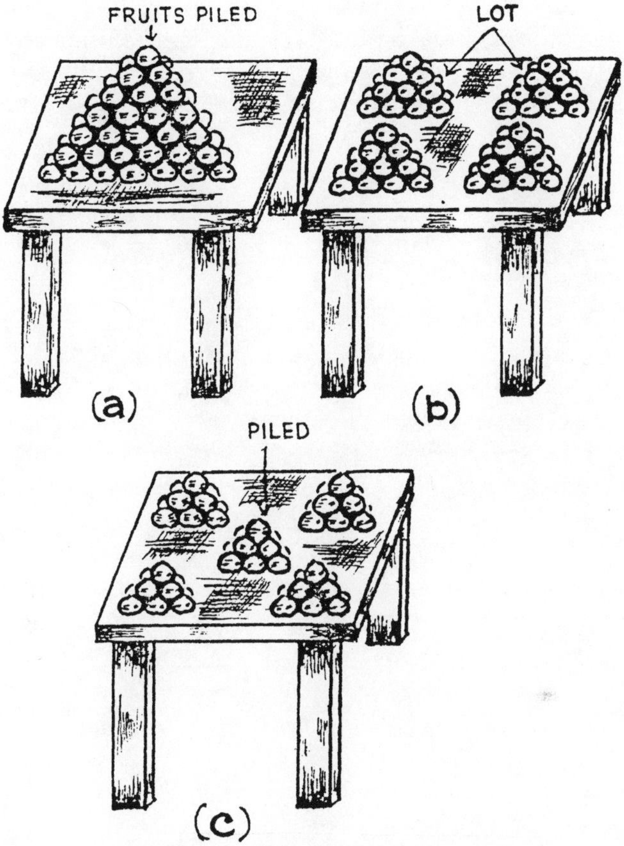

Fig. 4: Making analytical sample of fruits.
(a) Sorted fruits mixed together and piled. (b) Pile is broken to make several lots. (c) Some fruits are taken from each lot and again piled.

In order to make a representative sample of intact fruits, those which have been sorted out and surface-cleaned as described earlier under the headings, 1.3 and 1.4 should be used. These fruits should at first be thoroughly mixed together and piled at a place (Figure 4 a). Then this pile is disrupted and several lots are made with them at random basis (Figure 4 b). After that, some fruits are randomly taken from each lot and these are mixed together and heaped again (Figure 4 c). The remaining ones are discarded. This heap of fruits is broken again and the same procedure is followed. In this way, the process is repeated several times and ultimately, the representative sample of *intact fruits* is made. Evidently, more the process is repeated, more the correctness of the sample will be.

The above procedure has much suitability when the number of fruits is large. But where the fruits obtained are lesser in number, repeated sorting as stated above may be difficult to perform. Therefore, for practical purpose, one or two sortings may only be resorted to in such case.

An important point that should be remembered is, the total process starting from plucking of the fruits till the representative sample is made with them should be performed quickly for studying the physical or chemical status of them.

3

Determination of Constituents by Physical Methods

Physical components of fruits refer to those, for determination of which chemical procedure is not needed. For determining most of the physical components, intact fruits are used. However, in some cases like determination of juice content, pulp colour, inedible part, seed content etc., fruit samples are required to be cut.

There may in fact, be a large number of physical components of fruits that are worthy to test and the fruit analyst selects those which are of concern to him in the study undertaken. However, those that are more commonly considered in fruit analytical procedure have only been considered in the present text and described in the following pages.

3.1 Weight

The determination of weight of fruit samples needs no emphasis in any analytical study on fruits. Apart from estimating yield of a tree, weight of a fruit is considered to be an important factor in judging its compactness, maturity, juice content, levels of carbohydrate and other chemical constituents. The weight of a fruit also determines its acceptance to consumers and thereby, the market price of it.

The weight of a given sample of fruits can however, be determined easily in any standard physical balance. Various types of such balances are available (Figure 5), having one or two pans in each and are worked with by putting weights from outside or without those. In any case, the balance should be in proper condition without having any defect in it.

N.B. While expressing weight of any material, altitude and latitude of the place where its weight is taken need also be stated.

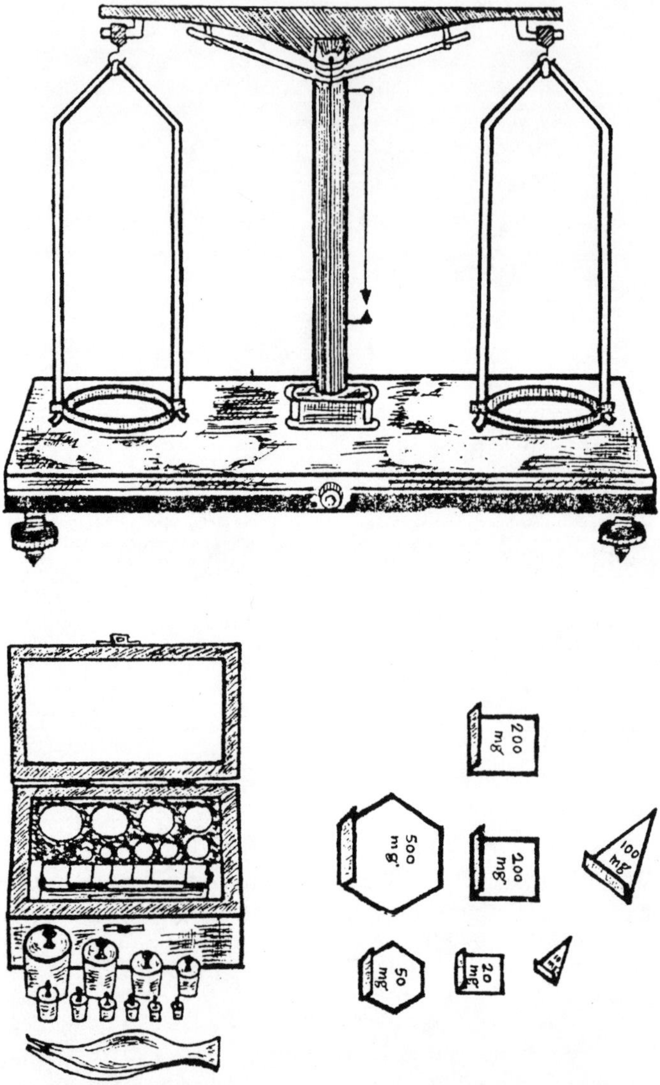

Fig. 5: Physical balance and weights.

To get the correct weight of the sample, the balance should be properly set, placed and levelled, accuracy ensured before use, weighing done accurately and the reading noted carefully. If necessary, weighing of the same sample may be repeated to make sure of it.

3.2 Volume

Volume, *i.e.*, size of a fruit is another important factor like that of its weight in determining quality. In the market, the consumers are also found to have a preference to select the large-sized ones for many fruits, *e.g.*, oranges, guava, peach, *jamun*, *aonla* and *bael*, *ber*, jackfruit, sapota etc. and accordingly, the price of those fruits goes higher with size.

The convenient method which is commonly used to determine the volume of a given fruit is by measuring the volume of a liquid, *i.e.*, water, which is actually displaced by it. This may be performed in a number of ways and some are described below.

Procedure

Measuring volume of displaced water directly in measuring cylinder

This is perhaps the simplest method to determine the volume of fruit samples. A measuring cylinder which is properly graduated and transparent should be used. Measuring cylinders of various capacities and diameters are in fact, available (Figure 6a) and among these, one in which, the sample of fruits can be easily and freely introduced should only be used.

Before use, the measuring cylinder should be thoroughly and repeatedly washed with clean and preferably running water. Then clean water should be poured in it. This should be done with care that no air-bubble gets trapped in the water that is poured and for this, the water should be poured slowly, holding the cylinder in a slanting manner. Nevertheless, if any air-bubble is noticed, it should be expelled out by striking the outer wall of the cylinder lightly with a stick or finger.

The water poured in the cylinder should be left to stand for sometime without disturbing it. After that, some water from it should be taken out by pipetting or siphoning to such an extent that the water left in the cylinder comes to a known volume and this should be known by careful observation of the lower meniscus of water layer coinciding with mark of the cylinder. The water layer in the cylinder should however, be sufficiently below the top (final) mark of it. The volume of water in the cylinder is noted and this is taken as the initial volume of water (Figure 6 b).

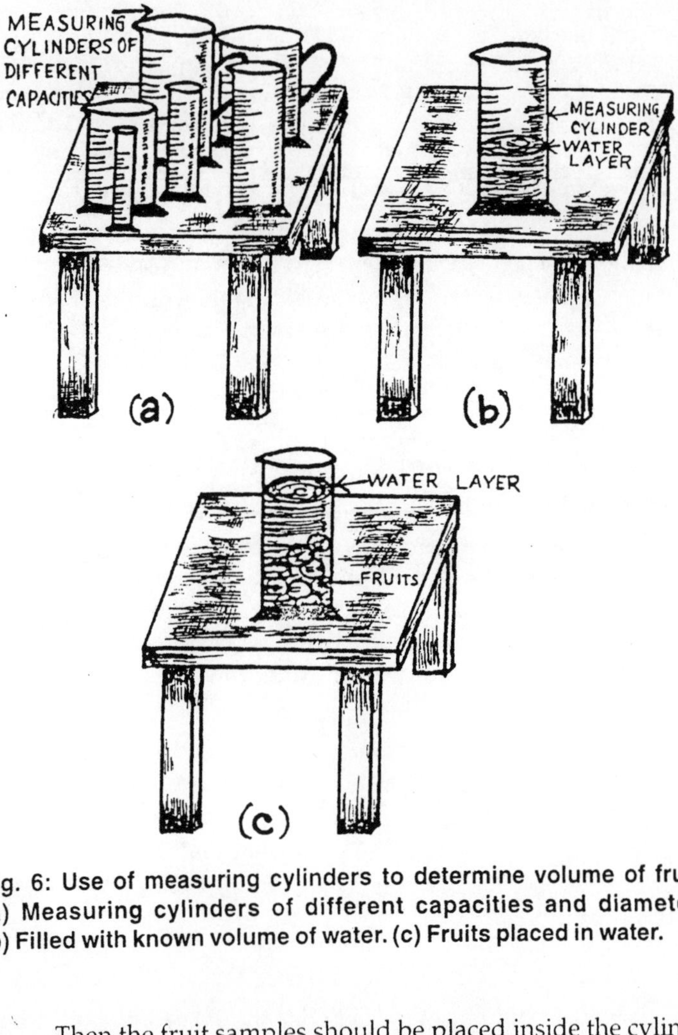

Fig. 6: Use of measuring cylinders to determine volume of fruits. (a) Measuring cylinders of different capacities and diameters. (b) Filled with known volume of water. (c) Fruits placed in water.

Then the fruit samples should be placed inside the cylinder containing water. This should be done slowly and very carefully such that the water does not splash and spill outside the cylinder. To do so, the cylinder may be held conveniently in a slanting manner. Care should also be taken that no air-bubble is trapped in the water while introducing the fruits.

When the fruits have sunk in water, they are left to stand for a while. If any air-bubble is found to be trapped in water, it should be driven out like before. Then the marking in the measuring cylinder coinciding with the lower meniscus of water is noted. This gives the final volume of water in the cylinder (Figure 6c).

After that, fruit samples are brought outside the measuring cylinder. The difference between the final and the initial volumes of water recorded in the cylinder gives the volume of the fruit samples tested with. This may be worked out as follows.

$$V = b - a$$

where,

V = Volume of all fruits present in the sample.

b = Final volume of water.

a = Initial volume of water.

Out of many fruits in the sample, the average volume of a fruit may be known by dividing the difference in volume with that of the number of fruits present in the sample as given below.

$$AV = \frac{b - a}{n}$$

where,

AV = Average volume of fruits present in the sample.

b = Final volume of water.

a = Initial volume of water.

n = Number of fruits present in the sample.

The above method is satisfactory if the fruits are heavier than water, *i.e.*, if they submerge completely in water. But if a fruit does not fully sink, the method as such cannot be used. In such case, as shown in Figure 7, the following device may be adopted.

A slab which fits snugly in the circular area of the measuring cylinder and is heavier than water such that it sinks in it easily should be taken (Figure 7a). It may be attached with a string with suitable arrangement, ensuring that it can be easily suspended by holding the thread. The slab should be of iron or of such material that is insoluble in water, not porous and thereby, it should not soak any water.

The measuring cylinder is filled with known volume of water (Figure 7b) in the same way as has been described above. The volume of water is noted and this gives the initial volume for the sample of fruits with the slab and say, this is *a*.

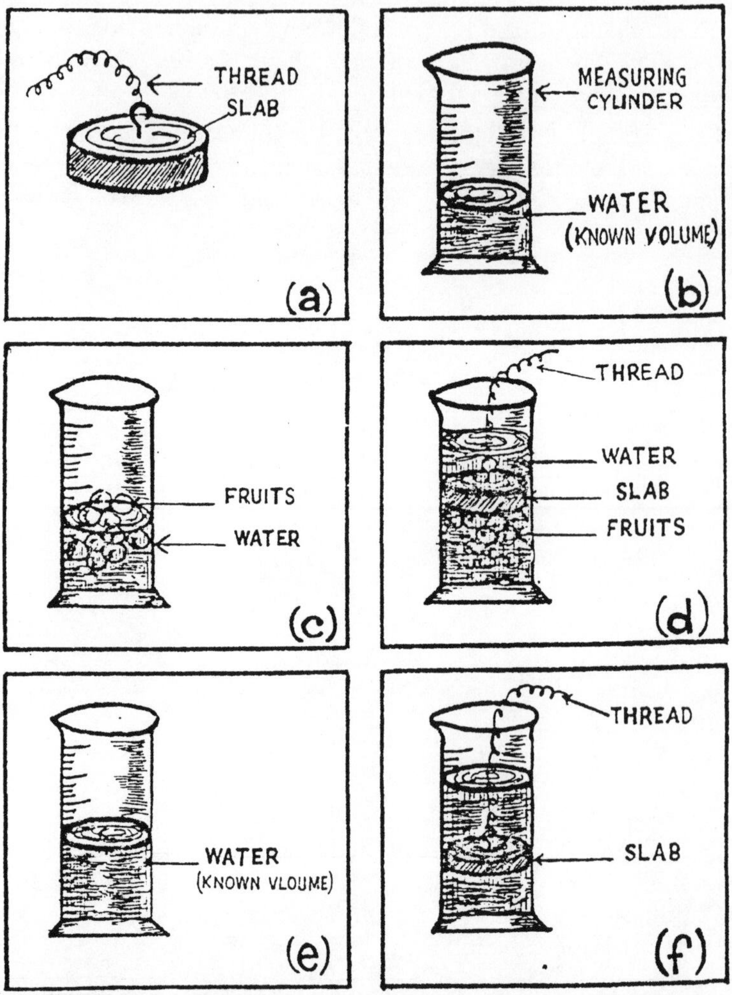

Fig. 7: Use of a slab to force water to sink and determining their volume. (a) Slab with thread. (b) Measuring cylinder filled with known volume of water. (c) Fruits placed in water. (d) Slab placed over fruits to sink them. (e) Measuring cylinder filled with known volume of water. (f) Slab placed in water.

Now, the sample of fruits are placed in the cylinder containing water, taking the same precautions as stated above. The fruits, either some or all of them may be seen not to sink in water (Figure 7c). Then the slab should be placed over the fruits in the cylinder slowly and very carefully. For convenience, the thread attached to the slab may be held to get it suspended and placed slowly over the fruits. The weight of the slab will force the fruits to sink into the water in the cylinder (Figure 7d). If any air-bubble is noticed inside, it should be removed out in the same way as stated earlier. The volume of water in the measuring cylinder is noted and this gives the final volume for the sample of fruits along with the slab and say, this is *b*.

After that, the slab and the sample of fruits are taken out from the cylinder. The cylinder is filled with water again as stated earlier (Figure 7e). The volume of water is noted and this gives the initial volume for the slab alone and, say this is *c*.

Then the slab is placed in the water. Air bubble, if any is noticed is driven out and the volume of water in the cylinder is noted (Figure 7f). This gives the final volume of water for the slab alone and say, this is *d*.

The slab is then brought out of the cylinder. Average volume of fruits in the sample may be determined as follows.

$$AV = \frac{(b-a)-(d-c)}{n}$$

where,

AV = Average volume of fruits present in the sample.
b = Final volume of water.
a = Initial volume of water (for b).
d = Final volume of water (for slab).
c = Initial volume of water (for d).
n = Number of fruits present in the sample.

Measuring volume of displaced water on transference to measuring cylinder

The method described above under (a) is easy, simple and gives results with much accuracy for such fruits which are of small size *e.g.*, grape, *ber*, *jamun*, *phalsa*, cherries, longan, some types of berries, *nor*, some varieties of *anola*, lime etc. It is also applicable for other fruits at earlier stages of this maturity, *i.e.*, when they are smaller in size.

On the other hand, for relatively large-sized fruits such as mango, guava, papaya, apple, pear and many other fruits, the above method has limitation, because these fruits are in fact, not possible to be introduced into a measuring cylinder. However, large-sized, graduated measuring cylinders having wide diameter are also available and these may be employed to measure volume of those fruits. But in a measuring cylinder of very wide diameter, there may be practical difficulty to read the meniscus of water layer in it with fair degree of accuracy, in other words, the volume of water that has actually been displaced by the fruits within the cylinder.

For fruits of such large size, it seems to be a better method to allow water to displace by introducing the fruit samples in a water-filled container at first and then to determine in a measuring cylinder the volume of water that has been displaced outside the container.

Some of the methods based on the above principle have been described below.

Method 1

In this method, a suitable container in which the fruit samples can be easily and snugly introduced and a tray over which the same container may be flatly rested upon leaving adequate space on all sides should be taken (Figure 8a).

Both the container and the tray should at first be thoroughly washed with water. The water adhered over their surfaces should then be wiped out with a piece of clean cloth or absorbent cotton. After that, these should be dried till water is completely effaced, especially from the outer surface of the container and inner surface of the tray. This may be done by rubbing with a cloth-piece or absorbent cotton.

When dried, the tray should be placed on a flat surface and over it the container should be placed in the centre.

The water is then continued to pour slowly into the container, preferably through a funnel or an inlet tube till it reaches upto brim. Utmost care should be taken to see that the container is filled fully to the brim yet no drop of water overflows from it and falls on the tray. To ensure this, the last part of water should be poured very slowly and even in drops.

When the container is filled with water to the brim (Figure 8b), the fruit samples should be slowly and carefully placed in it. On placing the fruits, the water will be displaced outside the container and evidently, this displaced water will get collected in the tray. The displacement will be complete when the fruits will sink completely (Figure 8c). However, due to less careful placement of

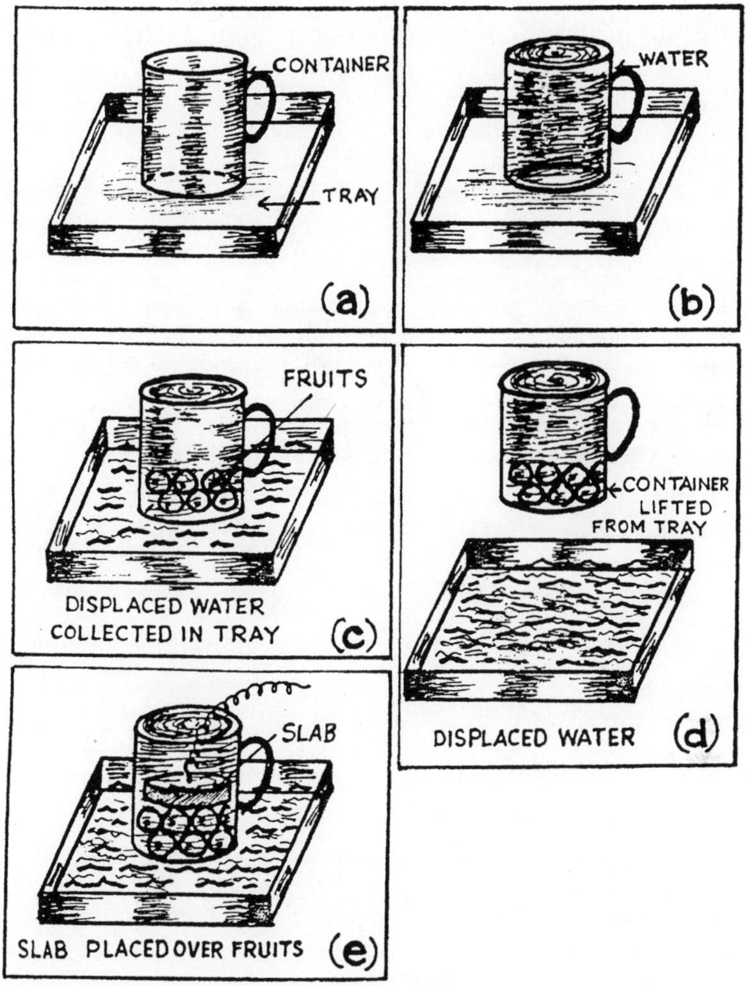

Fig. 8: Use of a container and tray to determine volume of fruits. (a) Container and tray. (b) Container filled with water to the brim. (c) Fruits placed in the water-filled containers and displaced water is collected in tray. (d) Container with fruits and water lifted from tray. (e) Slab placed over fruits in container to force them sink.

fruits, water may splash such that some extra water may go outside the container. In such case, little amount of water should be taken from the tray and this should be completely poured into the container. This will fill the container fully upto brim while the excess will overflow and collect in the tray.

Now, the water-filled container in which fruits are sunken is lifted out (Figure 8 d) with extreme care to ensure that no amount of water from it falls over the tray. Water if any, adhered over the outer surface of the container should also be taken in the tray. Lastly, the water displaced from the container and collected in the tray should be transferred to a measuring cylinder to determine its volume and this will be the same as the volume of the fruit samples. Average volume of a fruit belonged to the sample may be known on dividing the volume by the number of fruits.

If the fruits do not sink, *i.e.*, remain floated, a suitable slab of iron or heavy material as has been described above (Figure 7) may be used to force the fruits sink under the water (Figure 8e). When sunk, the volume of the displaced water is noted and this gives the volume of the fruit samples and the slab. After that, the volume of the slab alone should also be determined without use of fruits. From this, the volume of fruit samples alone may be known and the average volume per fruit may be worked out as follows.

$$AV = \frac{b - a}{n}$$

where,

AV = Average volume of fruits present in the sample.

b = Volume of all fruits present in the sample along with slab.

n = Number of fruits present in the sample.

The device described above has however, practical difficulty in its application. Some water is very likely to splash and fall outside the container, *i.e.*, in the tray when the former is filled with water upto the brim. In such case, it would become necessary to take out some water from the tray and return it to the container to replenish the loss. Besides this, removal of the water-filled container having fruits in it from the tray after displacement of water is also cumbersome. In this process of removal, some amount of water is likely to fall from the container into the tray and as this extra water is not required to be measured, the correct volume of water that has been actually displaced by the fruits cannot be known.

The above difficulties can be much overcome by adopting the methods 2 or 3 as presented below.

Method 2

In this method, an apparatus* consisting of 3 metallic components which do not rust are used.

One component consists of a cylindrical vessel of desired size having normal opening at one end and a funnel-like extension at the other end (Figure 9a). A crossbar is attached horizontally at the inner wall of the vessel (Figure 9b). The vessel is vertically erected with 3 stands fixed to it equidistantly.

The three vertical stands extend upwards above the vessel and their ends are welded together with a rim. The rim is also fixed with a horizontal crossbar as its spokes (Figure 9d). At the upper part of the interconnecting point of the two spokes is provided with a handle and just below it is suspended a rotating pulley (Figure 9e). The entire component is shown in Figure 9f.

The other component of the apparatus consists of a cylindrical box which is closely perforated at all sides. The box is provided with a perforated lid at the top that may be hinged on closing and a hook at the upper surface of the lid in its centre. The bottom of the perforated box is also provided with a heavy slab. The entire component is shown in Figure 9g.

The third component of the apparatus consists of an open-mouthed cask having a flat bottom, which is of size that can be freely introduced into the vessel (Figure 9h).

To work with the apparatus, the fruit samples are to be taken in the perforated box. It is then covered with the lid and the lid is hinged (Figure 9i). The hook provided at the upper part of the lid is tied with a strongly built thread or a flexible wire (Figure 9j). The thread or the wire is then passed through the groove of the pulley. It is pulled to lift the perforated box containing fruits upwards and close to the pulley. The box is allowed to suspend at this position by fastening the thread with the handle giving a knot that can be easily opened (Figure 9k).

Now, the cask is placed inside the vessel to rest it on the horizontal crossbar. Water is slowly poured into the cask through a funnel or an inlet tube till some water overflows from it, thus ensuring that it has been filled fully upto the brim. Care should be

* Available as Mazumdar's apparatus to detect volume of fruits.

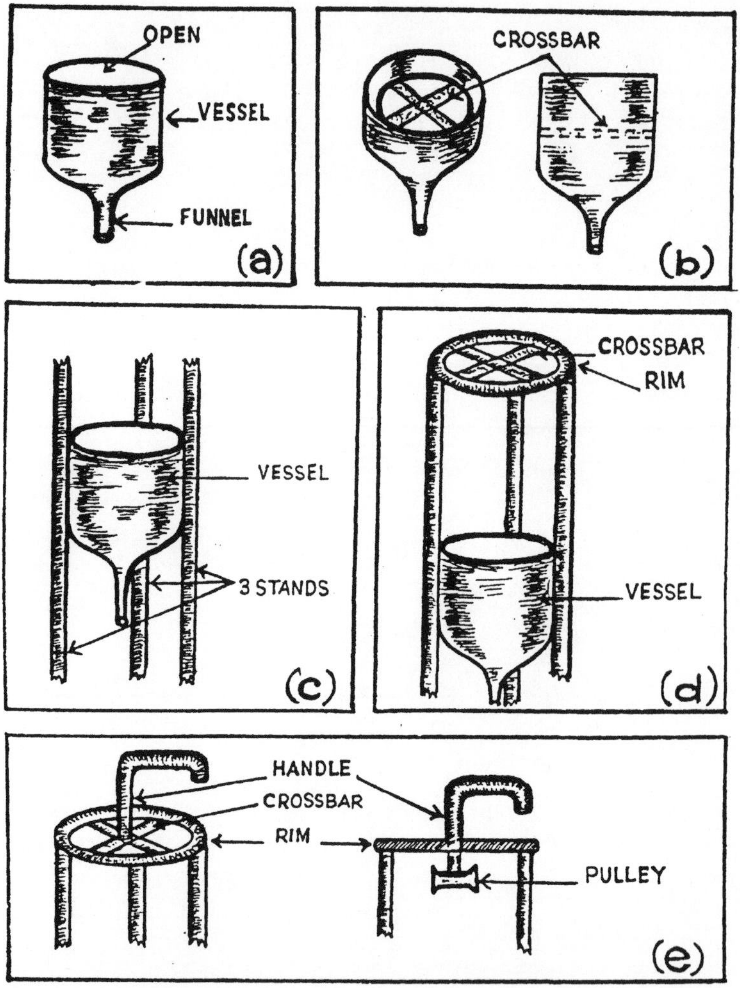

Fig. 9: Mazumdar's apparatus to detect volume of fruits.

(a) Cylindrical vessel with funnel end. (b) Crossbar attached at the inner wall of the vessel. (c) Vessel erected with 3 stands. (d) Rim with a crossbar at the top of the 3 stands. (e) Handle at the upper and pulley at the lower surfaces from the centre of the crossbar.

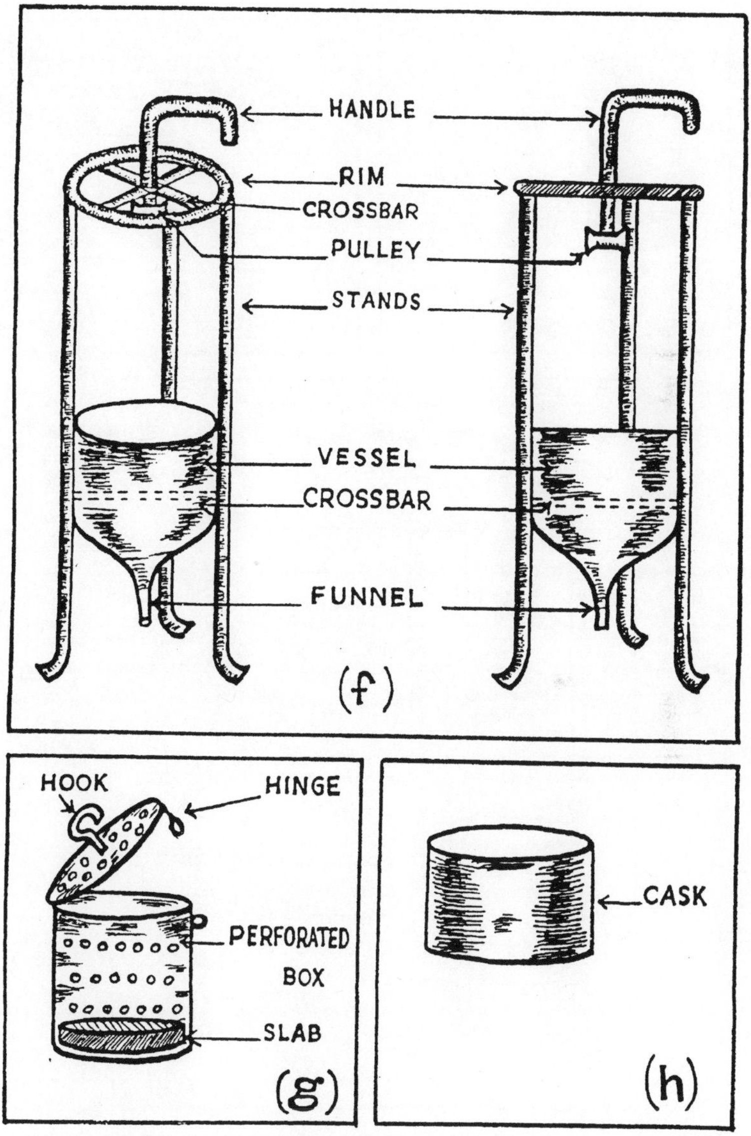

Fig. 9: Mazumdar's apparatus to detect volume of fruits.

(f) Entire component. (g) Perforated box with lid at top and slab below. (h) Cask.

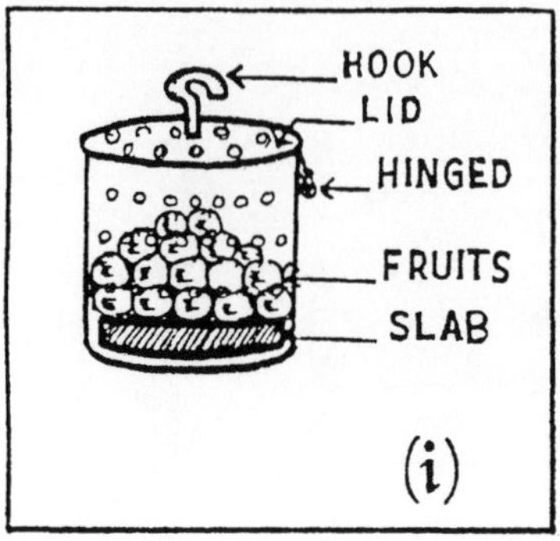

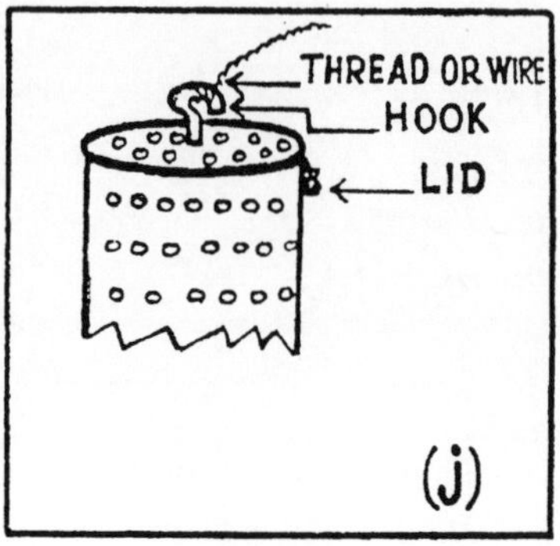

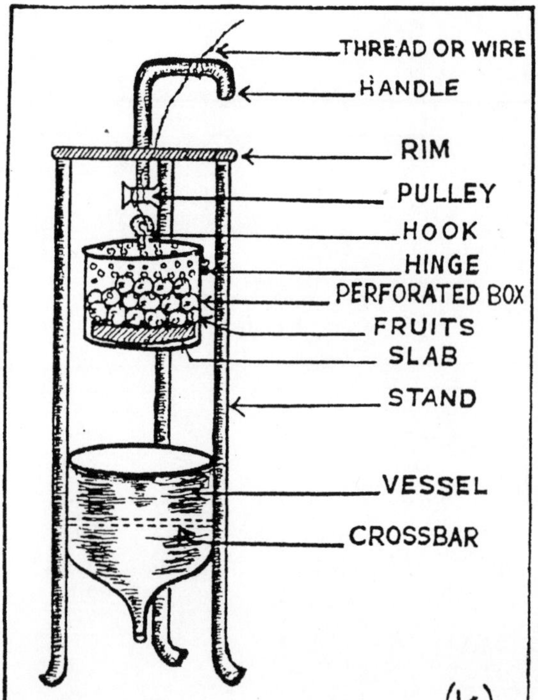

Fig. 9: Mazumdar's apparatus to detect volume of fruits.

(i) Fruits placed in perforated box, covered with lid and hinged. (j) Hook at the lid tied with thread or wire. (k) Perforated box pulled upwards and thread or wire knotted with handle.

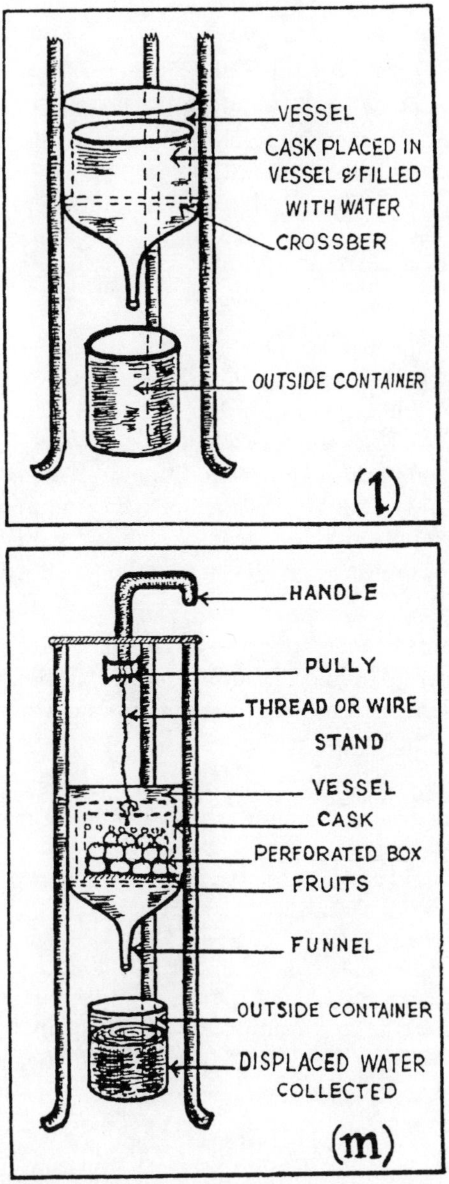

Fig. 9: Mazumdar's apparatus to detect volume of fruits.

(l) Cask placed inside vessel and filled with water to the brim. Container placed below funnel. (m) Perforated box rests on crossbar in the vessel by releasing the thread. Water displaced from cask is collected in outside container.

taken in pouring that no air-bubble is trapped in the water. A container is placed just below the funnel (Figure 9l).

Without disturbing the set, the thread fastened to the handle is unknotted and it is slowly released. By that, the perforated box containing fruits will move downwards into the cask till its bottom reaches the bottom of the cask. While the container goes on dipping, the water from the cask will be displaced neck and neck and in this way, the displacement of water will continue till the box is fully immersed, *i.e.*, till it rests on the cask.

Evidently, the displaced water overflowing from the cask will reach the wall of the vessel and will flow outside through the funnel end. This water will be let out in the container placed below the funnel (Figure 9m). The volume of the displaced water thus, collected in the outside container is transferred to a measuring cylinder to measure its volume. This volume equals to the volume of the fruits together with the perforated box and the slab. From this volume, the volume of the perforated box with slab is subtracted to get the volume of the fruit alone.

The volume of the perforated box with the slab placed at its bottom is usually written on it but in absence of that, this may be known by the same process of dipping it into the water in the cask without putting the fruits in it.

Having measured the volume of the displaced water, the average volume of the fruits in the sample may be determined as follows.

$$AV = \frac{b - a}{n}$$

where,

AV = Average volume of fruits present in the sample.

b = Volume of all fruits present in the sample, along with perforated box and slab.

a = Volume of perforated box with slab.

n = Number of fruits present in the sample.

Method 3

In this method, a simplified form of apparatus to displace water is used.

The apparatus* consists of two metallic components, one of which is the cylindrical vessel and the other, the perforated box. The vessel has open mouth above and a flat bottom below such that it can be rested on ground or on any flat surface. A few centimeters below the top, it is provided with an outlet pipe, which is bent downwards (Figure 10a).

From the two outer walls of the vessel, two vertical bars are fixed to it vis-a-vis and these are extended upwards (Figure 10b). The ends of the bars are joined together with a horizontal bar (Figure 10c). At the centre of the upper surface of the horizontal bar is provided with a handle and just below it at the lower surface, a rotating pulley is suspended The complete structure of the component is shown in Figure 10d.

The other component of the apparatus consists of a perforated box, which is exactly the same as is used in the former apparatus presented under method 1 (Figure 9g). Like that of the former apparatus, it also contains a heavy slab at the bottom, a lid with hinge at the top and a hook at the centre of the upper surface of the lid to which is tied a thread or a flexible wire.

To work with the apparatus, fruit samples are taken in the perforated box and then it is covered with the lid and hinged (Figure 9i). The thread or the wire which has been tied to the hook at the upper surface of the lid (Figure 9j) is passed through the groove of the pulley. On pulling the thread or the wire, the perforated box containing the fruits is lifted upwards close to the pulley and is suspended here by fastening the thread or the wire with the handle in a loose manner (Figure 9k).

Water is slowly poured into the vessel through a funnel taking care that no air-bubble is trapped. Pouring of water is continued till some amount of it is discharged outside through the outlet pipe. When discharge of water has ceased, a container is placed below the outlet pipe (Figure 10 e).

Then the thread or the wire is unknotted and the perforated box containing the fruits slowly released so that it moves downwards (Figure 10 f). It is released further till the box sinks into the water kept in the vessel and ultimately rests at the bottom of the vessel. As the box goes on dipping, the water from the vessel will also be

* Mazumdar's modified apparatus to detect volume of fruits.

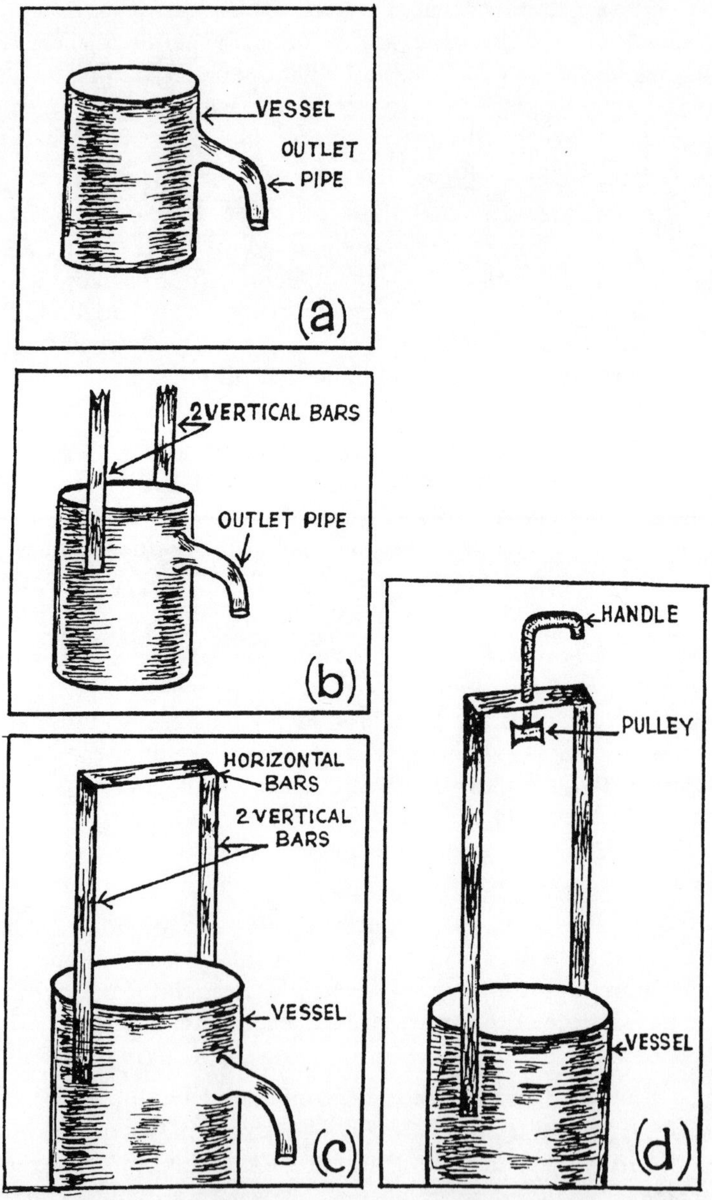

Fig. 10: Mazumdar's modified apparatus to detect volume of fruits. (a) Cylindrical vessel with outlet pipe. (b) 2 vertical bars are fixed at the outer wall of the vessel. (c) Horizontal bar at the top of the 2 stands. (d) Handle at the upper and pulley at the lower surfaces from the centre of the horizontal bar.

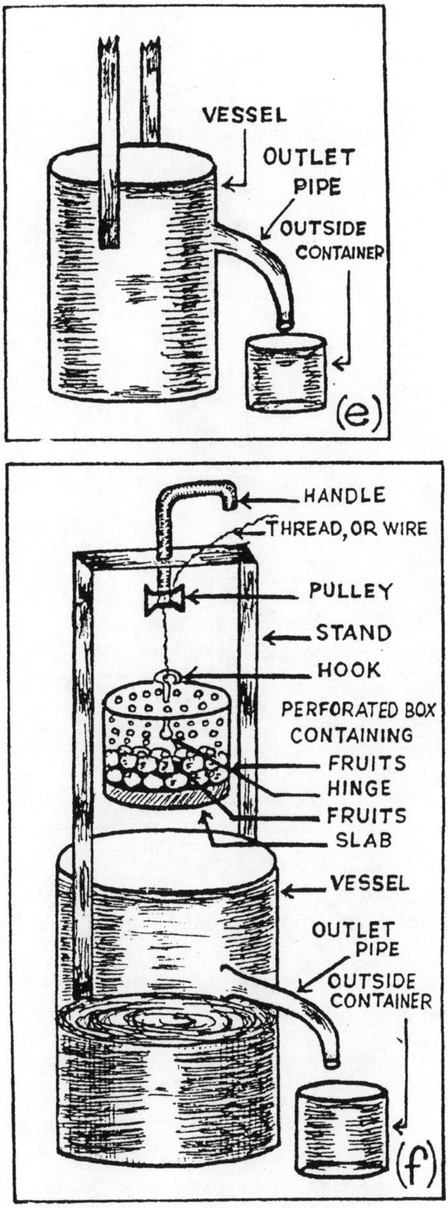

Fig. 10: Mazumdar's modified apparatus to detect volume of fruits. (e) Container placed below the outlet. (f) Perforated box containing the fruits is slowly released.

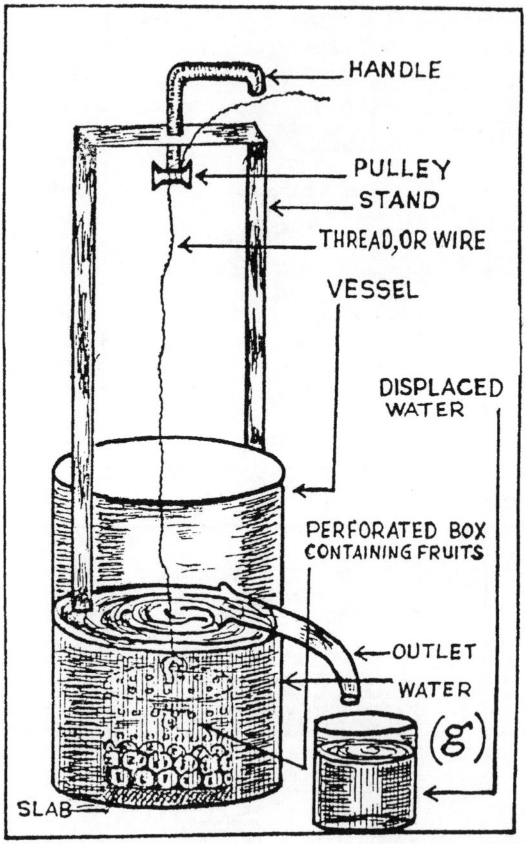

Fig. 10: Mazumdar's modified apparatus to detect volume of fruits. (g) Perforated box rests at the bottom of the vessel. Water displaced from the vessel is collected in the outside container.

displaced and will be collected in the outside container. The displacement will be completed till the box rests at the bottom of the vessel (Figure 10g).

The displaced water collected in the outside container is then transferred to a measuring cylinder to determine its volume. This volume gives the volume of the fruits along with the volume of the perforated box and the slab. From this volume, the volume of the fruits alone may be determined on subtracting the volume of the box with slab given on its surface, in absence of which however, this has to be determined following the same method of dipping it without enclosing the fruits.

Average volume of fruits present in the sample may be determined as follows.

$$AV = \frac{b - a}{n}$$

where,

AV = Average volume of fruits present in the sample.

b = Volume of all fruits present in the sample, along with perforated box and slab.

a = Volume of perforated box with slab.

n = Number of fruits present in the sample.

Other Methods

Among other methods to determine volume of fruits, hydrostatic balance (Figure 11) may also be used, although there are practical difficulties in using it and when immersed, the fruits should not touch the walls of the container.

Considering diameter of a fruit

For such a fruit, which is *spherical* in shape as is found in some varieties of sapota, elephant apple, lime, *aonla*, *ber*, plum, guava, grape, cape-gooseberry, etc. the volume may also be determined by considering its radius (half diameter) and adopting the following formule:

$$V = \pi \times r^3 \times \frac{4}{3}$$

where,

V = Volume of the fruit.

r = Radius of the fruit.

(That is, multiplying the cube of the radius of the fruit by 3.1417, multiplying by 4 and dividing by 3).

However, it should be noted that accurate result is difficult to be obtained by the above formula as absolutely concentric spherical fruits are rarely noticeable.

3.3 Specific Gravity

The relative heaviness, measured as *specific gravity* value is an important character of a fruit. In many fruits, the specific gravity

N.B. While expressing volume of any material, its temperature at the time of determining the volume need also be stated.

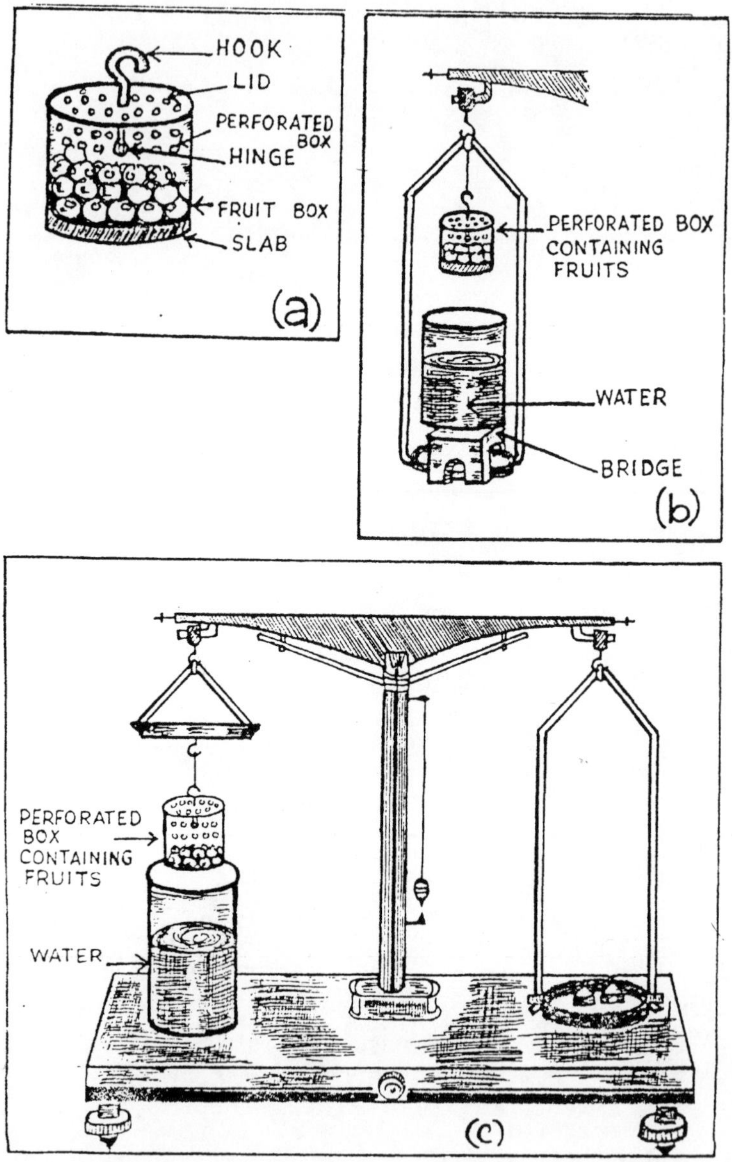

Fig. 11: Hydrostatic Balance. (a) Perforated box containing fruits. (b) Ordinary balance converted into a hydrostatic one. (c) Hydrostatic one.

value is considered to be a guiding index to judge their physiological maturity.

In citrus fruits, higher specific gravity value is closely associated with higher juice content and freedom from granulation of juice. Fermentation in the pulp as seen in pine-apple and some other fruits, the specific gravity value is related to season of their development, maturation and ripening. Use of any particular rootstock, as well as cultural conditions and nutritional status of the tree may also be responsible for specific gravity value of the fruits.

The alteration in the specific gravity value of fruits may also result on seed content of many fruits like jackfruit, custard apple etc., volume of the central hollowness as is found in fruits of some varieties of papaya, mandarin orange etc., sponginess of pulp and on other factors. Infection in the pulp as in *bael*, elephant apple and other fruits, attack by the insect, fruit-fly etc. are also known to bring about lowering of the specific gravity value in them.

The specific gravity value of fruit samples may be determined by the following methods.

Procedure

From weight and volume

Having determined the weight and volume of a fruit as described earlier (Ref. 3.1 and 3.2), its specific gravity may be determined as follows.

$$\textit{Specific gravity value} = \frac{\text{Weight of the fruit}}{\text{Volume of the fruit}}$$

Other methods

The specific gravity of fruits may also be determined with the help of a hydrometer or a hydrostatic balance (Figure 11).

3.4 Overall Length

An important qualitative character of a fruit is the length that it measures. For fruits of some species, maturity for harvest is also

N.B. While expressing specific gravity of any material, its temperature at the time of determining the volume, besides altitude and latitude of the place where the weight is taken need also be stated.

judged in considering the length attained by them. In the market, the consumers are even found to select out relatively long-sized fruits especially for those of banana, jack-fruit, pine-apple, lemon, papaya*, grape, *jamun* etc., although it is not certain whether a longer fruit has any impact on its quality.

The length of a fruit is referred to by many as the space, *i.e.*, straight line distance (displacement) between its stalk-end and the stylar-end. But it needs pointing out that the consumers consider the total length of the fruit without regard to the stalk-stylar length of it. Thus, it appears to be more appropriate to consider the total length of a fruit, which may be termed as its *overall length.*

By overall length of a fruit is therefore, meant the straight line distance spaced from the uppermost point at its apex and upto the lowermost of it at the base. It may be noted that although in a number of fruits like many varieties of papaya, lemon, guava, sweet orange, apple, peach etc. the overall length is the same as the stalk-stylar length, it is not so for all fruits. Mango is an example, the ventral shoulder in fruits of many varieties of which being highly raised, the overall length would in fact, be greater than the stalk-stylar length. The overall length of a fruit may be measured as follows.

Procedure

By the use of a slide-callipers

The overall length of a given fruit can be easily measured with the help of a slide-callipers (Figure 12), provided that it is of such length as can be suitably held between the jaws of the callipers. To measure the overall length, care should however, be taken to hold the fruit in correct position, ensuring that its uppermost and the lowermost ends are in contact, respectively with the two jaws of the instrument.

By the use of 'L'-shaped sets

A slide-callipers has limitations to measure the length of fairly long-sized fruits like mature papaya, *bael*, shaddock, pine-apple, jack-fruit etc. For such fruits and also those of relatively short length, the following method may be conveniently adopted, which employs two 'L'-shaped sets.

* In papaya however, spherical fruits, which are usually developed from female flowers are sweeter than the elongated ones developed from bisexual flowers.

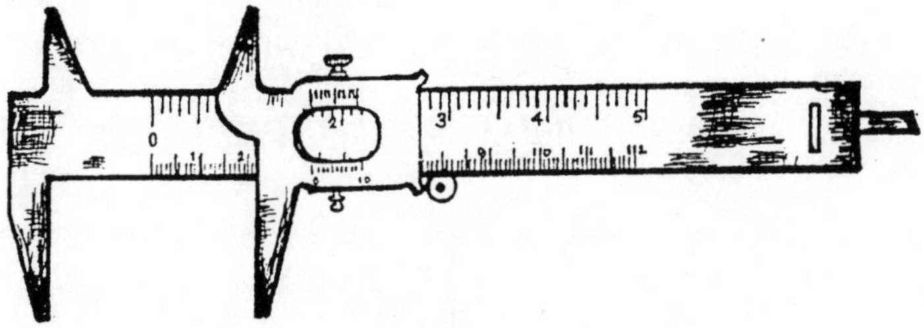

Fig. 12: Slide-callipers.

Two 'L'-shaped sets, which are vertically raised to few centimetres, have 90° angle (Figure 13a) and made up of any hard material such as metal, wood, plastic, cardboard should be taken. The length of the arms of each 'L' and its heights may vary according to the length and size of the fruit to be measured.

One of the 'L' sets should be placed on a paper kept over a flat surface and two lines to be drawn along two arms of it. Now the fruit, the length of which is to be measured should be brought close to it in such a position that one end of it touches the arm of the 'L' on which, the line has been drawn (Figure 13b).

Without disturbing the position of the fruit, the arm of the second set of 'L' should be brought in contact with the corresponding arm of the first set of 'L'. The former 'L' set is then slowly drawn downwards until it touches the other end of the fruit (Figure 13c). At this position, another line should be drawn along the arm of the second set of 'L'. After drawing this second line, which will be parallel to that of the first line, the 'L' sets and the fruits should be removed. The distance between the two parallel lines is then measured which will be the overall length of the fruit.

If the fruit tends to roll and thereby, its position is liable to be changed, it can be placed over the open side of a suitable lid to maintain its stationary position (Figure 13d).

3.5 Maximum Width

The width of a fruit, *i.e.*, how much it has grown across is a measure which is perhaps of greater importance than the lengthwise growth of it, in order to judge its normal and healthy growth. In some countries, the maturity for harvest of many fruits is also assessed by observing their girthwise growth. For example, in the fruits of limes and lemons, width is sometimes considered to determine their harvest-maturity condition. The practice lies in testing whether the fruit passes or not through a ring of desired diameter. If the fruit passes out, it is considered not to have fully grown till then (Figure 14a), for which it is left to grow further *in-situ*,

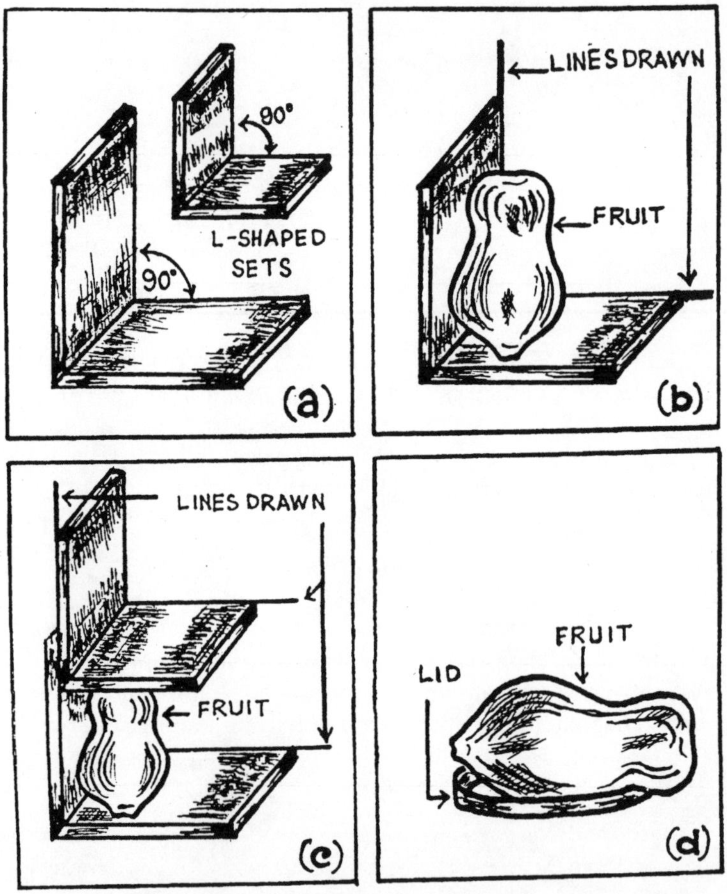

Fig. 13: Measuring overall legth of a fruit. (a) Two 'L'-shaped sets of hard material. (b) One end of the fruit touches one arm of the 'L'. Lines are drawn along the arms of the 'L'. (c) Arm of the other 'L' set is touched with other end of the fruit. Line is drawn along the arm of the 'L'. (d) Fruit is placed over open side of a lid to hold it in position.

N.B. While expressing length of a fruit, its temperature at the time of measuring need also be stated.

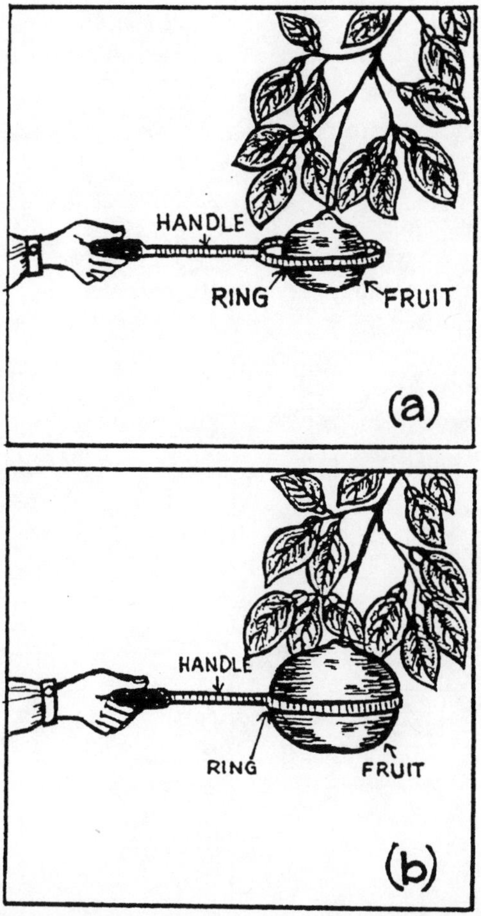

Fig. 14: Testing harvest-maturity of a spherical shaped fruit using a ring. (a) The fruit passes through the ring and is not considered ready for harvest. (b) The fruit does not pass through the rind and is judged to have attained harvest-maturity.

instead of harvesting at that stage. On the other hand, if it fails to pass through the ring (Figure 14 b), it is considered to have grown to desired degree and hence, it is harvested then.

To refer to width, the diameter of a fruit in its centre is emphasized by many fruit analysts. In fact, in some fruits which are spherical or oblate in shape in particular, the cross-sectional diameter is the same as the maximum width. However, this is not the case for all fruits and therefore, many analysts are in favour of considering the total width of a fruit, measuring the distance from the extreme points at two sides.

The width of a fruit is measured in similar way as its length is measured (Ref. 3.4).

Procedure

For such a fruit which can be held crosswise between the two jaws of a slide-callipers (Figure 12), the method can be used satisfactorily. In large-sized fruits however, the 'L'-shaped sets as have been used to measure the length (Figure 13) may be conveniently used. Unlike in case of length, the fruit should be held across between the arms of two 'L' sets (Figure 15) when the maximum width of it is needed to be determined.

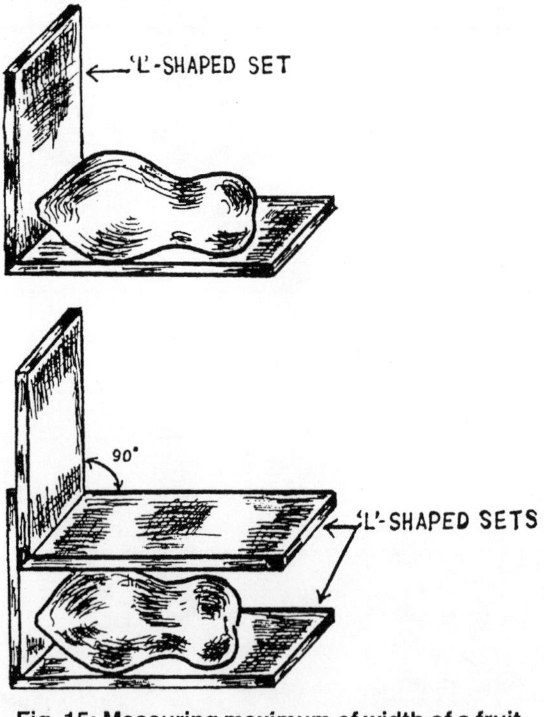

Fig. 15: Measuring maximum of width of a fruit, using two 'L'-shaped sets.

N.B. While expressing width of a fruit, its temperature at the time of measruing need also be stated.

3.6 Shape

Fruits belonged to a species or a variety of it have some characteristic shape of their own, although variation within some limit is not considered to be an uncommon feature. The fruits commonly consumable to the human beings are found to have number of shapes (Figure 16) and a mixture of more than one shape may also exist in them.

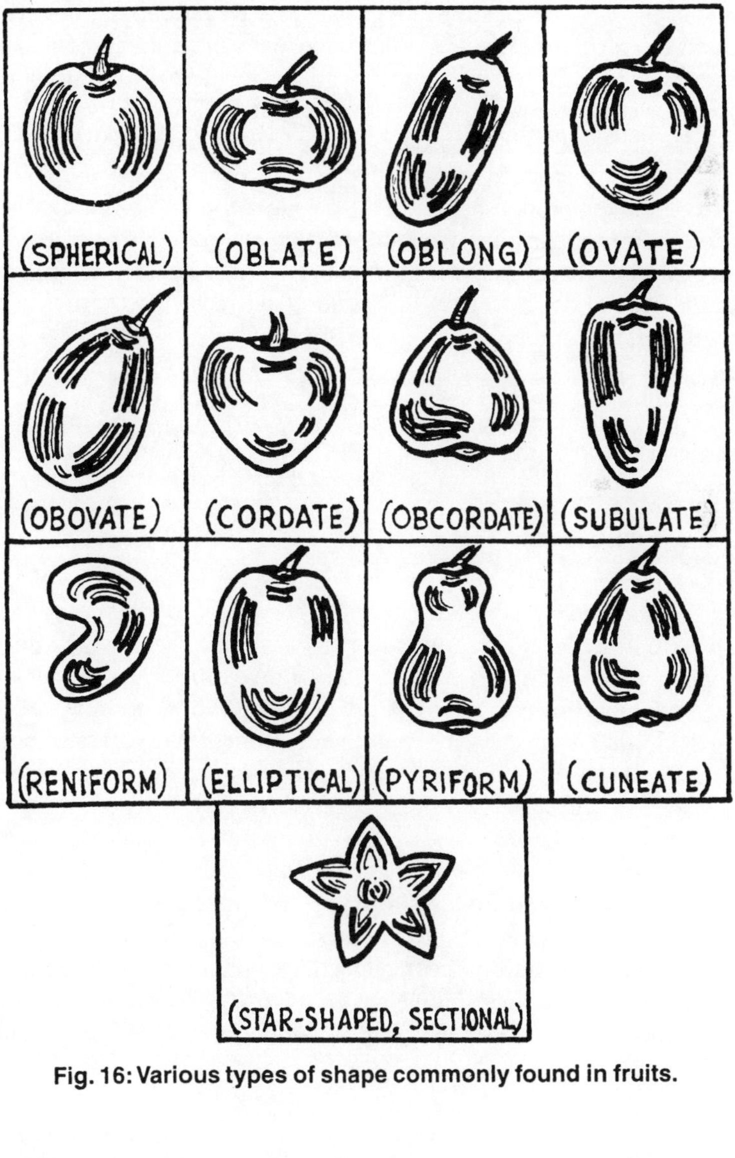

Fig. 16: Various types of shape commonly found in fruits.

The study of the shape as a physical feature of a fruit has its importance to determine whether it conforms to the standard and any deviation that may result due to interacting factors like environment, biotic agencies or application of agro-inputs to the trees bearing them.

Modification in the shape of a fruit due to depression taking place at one or more sites of it is sometimes met with in some fruits and especially in the many-seeded ones, *e.g.*, custard apple, jackfruit etc. The demerit is usually associated with production of lesser number of seeds in the fruit as a result of insufficient pollination. Experiments in custard apple have produced evidence that the berries lying at the depressed part of the fruit which are seedless are qualitatively inferior than the seeded berries. The importance of seeds in maintaining proper shape of fruits belonged to a number of species, though not all has been well-substantiated. A notable example is strawberry, where removal of the *achenes* from any part of the fruit at early stage of development brings about a spectacular depression on that part.

Procedure

Visual observation is beyond doubt, the easiest and practical method to judge the shape of a fruit and any deviation from the norm.

3.7 Firmness

Firmness of a healthy fruit is linked to the degree of its *physiological maturation*. With progress of development, maturation and ripening either in the pre-harvest or in the post-harvest condition, the fruits undergo gradual softness to a greater or a lesser extent depending on species, varieties, environment and the use of agro-inputs. Chemically, the change-over from firm to soft condition of a fruit is largely due to enzymic conversion of *pectic compounds* present in them as component of cell-walls. These complex polysaccharides, which are long-chain polymers of galacturonate monomers gradually transform from water-insoluble to the more and more soluble forms, *i.e.*, from protopectin to pectic acid in the fruits as they undergo development, maturation and ripening. Due to this solubilization process of pectic polymers, the fruits attain more and more softness. However, the intensity of softness is not the same for the fruits under all species and those in which the degree of softness is relatively less, the magnitude of pectic catabolism eventuating solubilization of the polymers is also lesser.

The extent of firmness of a fruit may be determined by the following methods.

Procedure

By the use of penterometer

Firmness may be measured by this instrument with fair degree of accuracy.

By hand-feeling

An easy approach to judge firmness of a fruit is to hold it in palm and feel it by giving the desired pressure. The method is crude but practical and may be applied to have gross idea, especially when relative degree of firmness among different samples of fruits is needed to be assessed.

In order to lessen as much as possible the tester's personal factor in judging by hand-feeling, the test should not be carried by one person only. Instead of that, a *panel of persons* should in fact, be employed for the purpose. The judges should be adult, comprise both men and women and have normal strength in wrist. There should be several groups each consisting of two or more judges and each group should be assigned to test independently by the hand-feeling process.

The degree of firmness or softness as assessed by each group may be expressed by suitable terms such as very hard, medium hard, hard-to-soft, moderately soft, soft, soft-to-very-soft, extremely soft etc., which should be informed to the tester before starting the test. However, this may also be expressed in other ways also, such as by putting number of + and – signs to denote the intensity of firmness or softness respectively, or by marking with numerical value and the like. The reports obtained from each group of tester should be considered together and made average to get the inference.

By pin-drop method

This is also a crude method and may be of use to determine the relative level of softness of fruits. The method however, has little applicability for hard fruits and may be used for those fruits which are soft or moderately soft in texture.

To work with, a known weight, say from 0.5 g to 3 g or more as desired by the analyst should be attached at the head of a suitable pin (Figure 17a). The length of the pin should also be known and may be 1 cm or more.

Now, the pin with the weight above should be made to drop vertically and freely over the surface of the given fruit from a known height which may be 1 cm or more (Figure 17b). In containing the weighty head, the pin will pierce the fruit and go upto some distance (Figure 17c) depending on the texture of the fruit. The pin is then taken out and the length of it that has penetrated into the fruit is measured. The interacting effect of air on free drop of the pin may be averted by carrying out the test in an enclosed chamber. In absence of that, the test may be performed in a suitable box or chamber to avoid flowing air coming in contact with the needle.

To express the degree of softness of a fruit determined in this way, the procedural details is necessary to be given stating the length of the pin, the weight provided at its head, the distance from which it was dropped and whether the test has been carried out in an enclosed chamber or open condition.

3.8 Peel Colour

The surface colour of a fruit is an important qualitative character which has significance on many aspects and the pertinent ones have been stated below.

1. In fruits of many species, the change-over in colour of peel is associated with their maturation and ripening, either in pre-harvest or in the post-harvest condition of them. The onset of ripening in fruits of some species is marked by appearance of yellowish patch at some part on their green surface, which is known as *colour-break* and is evinced in many varieties of mango, papaya, sweet orange, shaddock, grapefruit, water-melon, pine-apple etc.
2. The colour of fruit surface is beyond doubt, an important factor in determining appeal to the consumers.
3. Attainment of deep red colouration in the fruit peel of some varieties of apple, mango, plum etc., which is primarily due to higher accumulation of anthocyanin pigment is indicative of the duration and amount of insolation received by them during development. The amount and duration

Fig. 17: Judging softness of fruits by pin-drop method. (a) Pin with a weight at its head. (b) It is dropped vertically. (c) Penetration of it in the fruit.

of cold exposure received also determine the intensity of red colour on the surface of some fruits as in strawberry, cherry, some varieties of guava and notably, with the *Apple Colour* variety of guava.

4. The soil *p*H, rootstock over which the tree is grown and the agro-inputs used are also factors affecting peel colour

of some fruits.

5. The characteristic change in the peel colour of some fruits may also be the symptoms of deficiency or excess of some plant nutrients.
6. Infection by some disease in the fruits and especially due to virus may also bring about change on the peel colour of them.

The colour of fruit peel is chemically due to accumulation of one or more forms of pigments in different combinations. Some of these pigments are soluble in water while some in fat or fat-solvents and these can be separated out and determined by chemical analyses. However, without going into chemical procedure, the relative level of pigmentation of the colour may be known in a gross manner by the following methods.

Procedure

By the use of a colour-dictionary

In this method, the peel colour of a fruit is compared with the chart provided in a colour-dictionary. The peel colour that conforms most closely to the specimen in the chart is noted and the colour is narrated in accordance with the description provided in the chart. Evidently, the colour-dictionary which has been used for comparison should be a standard and universally acceptable one. It should also be remembered that the description of colour should be given along with citation of reference of the colour-dictionary that has been consulted for the purpose.

Instead of maintaining uniform colour over the entire surface of the fruit, if more than one colour is persistent at different parts of it, comparison should be made separately for all these parts. For this purpose, as given in Figure 18, the entire fruit may be observed under separate parts and the colour observed at any such part should be narrated separately.

Eye-estimation

If colour chart is not available, colour of a fruit surface may be judged by eye-estimation also which is to be done by testers. But this way of determining colour does not appear to be more suitable than making a comparison with a standard colour-dictionary.

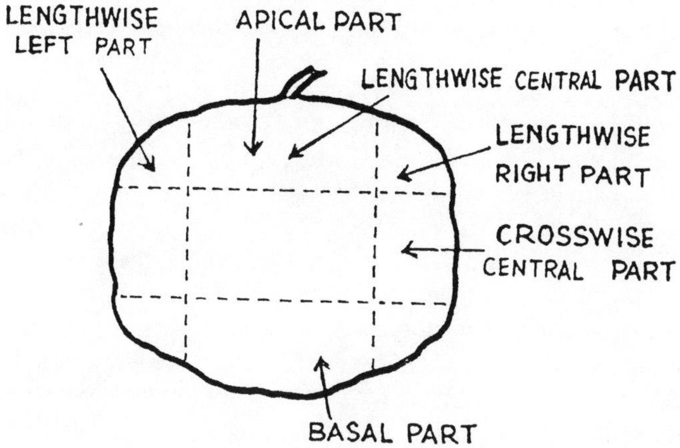

Fig. 18: Conveniently made parts of a fruit.

Judging the colour of a fruit should however, be done by one person only and in order to disregard one's own individual factor, if any, it should be made to conduct by a panel of persons as stated earlier while determining firmness (Ref. 3.7). Several groups of judges should be formed comprising men and women having vision in them and without colour-blindness in particular. Each group comprising two or more judges should be assigned to identify the colour as they observe and narrate the colour under specific terms which should be informed and explained to them before they start judging. The inference should be drawn making average of the reports given by the groups of judges.

3.9 Peel Smoothness

The peels of fruits belonged to many species and varieties are felt to get smoothened with advancement of their development, maturation and ripening. Such changes are noticeable in banana *fingers*, where the angulated ridges get flattened, in litchi and jackfruit, where the pointed ends of the tubercles tend to become obtuse, in pine-apple, where the eyes become less raised, in some guavas and citruses, where the warts become less prominent and the like. Paying heed to this character, the degree of smoothness attained by the surface of a fruit is among other indices considered to be a dependable one to judge harvest-maturity of those fruits.

In order to assess the degree of smoothness achieved by a fruit, hand-feeling if properly done, appears to be a practical method.

Procedure

On feeling the surface of a fruit with palm, the degree of smoothness attained by it may be known by an expert. The tester should however, have a knowledge on the extent of smoothness that is attained by the fruits of the variety under the species at different stages of their development when grown in the same locality. Only on having this experience, the tester can estimate the degree of smoothness attained by the given sample of fruits. As in case of determining firmness (Ref. 3.6) and peel colour (Ref. 3.7), smoothness of peel should however, be tested by a panel of testers and not by one person only.

It should also be remembered that surface-smoothness test should be performed for fruits of only those species and varieties where detectable change in smoothness is noticeable during their development, maturation or ripening and not for all.

3.10 Peel Wax

In many plant species, the leaves, stems, flowers and fruits are found to be coated with a thin layer of wax, known as *plant wax*. The deposition of wax is noticeable over the surface of many human consumable fruits also. The waxes in fact, originate as droplets of oil in the epidermal cells and passing through fine canals penetrating the cell-wall, deposit on the outer surface crystallizing into rods or platelets.

The physiological function of plant waxes though remains to be known in greater depth, their effect in retarding loss of water from the organs and thereby, protection to them gains support apart from possible hormonal influence of them. However, when deposited over the outer surface of a fruit, the wax is known to exert a beneficial role to protect it from adverse condition and especially, to enhance its shelf-life when stored after harvest. To derive this advantage of waxes, now-a-days these are commercially extracted from sugarcane trash or other waste materials where they exist in relatively high amount and are used to provide a coating over the surface of many fruits before their storage.

Chemically, a plant wax is an ester of a higher fatty acid with a long-chain alcohol other than glycerol. Besides the simple esters, the waxes are also known to contain alkanes of odd carbon numbers (C_{21} to C_{37}), primary alcohols with even numbers of carbon atoms (C_{22} to C_{32}) and long-chain free-fatty acids (C_{14} to C_{34}). The wax coated over a fruit surface may be extracted and estimated by chemical procedure. However, as stated below, the relative level of deposition of wax over the surfaces of different samples of fruits or at different stages of their development may be known in a gross manner by hand-feeling.

Procedure

To have a general idea on the extent of wax deposited over the surface of a given fruit, a simple way may be to rub the surface lightly with fingers or a piece of tissue paper. A reasonable amount of wax covered by a fruit surface may be known by an experienced tester and this may be expressed as high, medium or low on making suitable comparison. As in case of firmness (Ref. 3.6), colour (Ref. 3.7) and smoothness (Ref. 3.8), the waxiness of a fruit surface should also be got to be tested by a panel of persons rather than an individual tester only.

3.11 Peel Thickness

The thickness of peel is considered to be a character of importance in many fruits. In many citruses and also in pomegranate, the juice content of fruits is estimated to be higher with thinner the peels of them. Again, the thinner peel in the fruits of the *Langra* and some other varieties of mango has been a causative factor for their relatively low shelf-life while the fruits of *Alfonso* and other varieties keep well for longer time in having thicker peels over them. Same is true with many varieties of banana, sapota, litchi and other fruits as well. Higher thickness of peel may also facilitate handling, grading and packing of many fruits without rendering damage to them. In fruit analytical study therefore, determination of the thickness of peel is given emphasis for many fruits.

Procedure

For measuring the peel thickness, the first step is to excoriate it from the fruit. In case of fruits like mandarin orange, banana, litchi etc., peeling may be done easily by pulling it apart. For fruits of other group like mango, sapota, some citruses, custard apple, longan, pomegranate, *ber* etc. the peel is semi-tightly attached to the inner part and when flayed from the fruit, part of the pulp gets attached to it. Hence, after taking out the peel from such a fruit, it should be made free from the pulp adhered to its under-surface by rubbing with a blunt knife. Again, there are other fruits like sweet orange, shaddock, grapefruit, some other citruses, papaya etc., in which cases, the peel is more tightly attached to the pulp. Accordingly, it should be carefully cut out with a knife in order to separate from the inner part of the fruit.

Besides the above, some fruits, *e.g.*, apple, guava, grape, *jamun*, star-apple etc., are there in which the pulp is extremely thin and is so tightly attached to the pulp that there is practical difficulty to strip it off ordinarily by mechanical means. Measuring thickness of the peel of such fruits after flaying is in fact, difficult.

However, after the peel is stripped off from the fruit and if necessary, made free from adhering pulp, the thickness of it is measured. Evidently, this can be done easily and in a best possible manner with the help of a slide-callipers (Figure 12).

3.12 Peel Oil Gland

The oil-glands located in the peel of a fruit is given emphasis in analytical study of many fruits as they detemine flavour of them. The relative abundance of oil-glands in the peels is a notable character in many species and varieties of fruits. This also depends on other factors like type of rootstock used, edapho-climatic conditions of growing, agro-inputs that are applied notably, the plant nutrients and the growth-regulants. As compared to other fruits, in many citruses in particular, it is relatively easy to determine the intensity of oil-glands on a given area of the peel by counting as described below.

Procedure

The peel of the citrus fruit, *e.g.*, mandarin orange is at first stripped off (by pulling). Then the whitish inner layer of it (albedo) is removed which may be done by scraping with a scalpel. Scraping should be done carefully until the outer part of the peel (flavedo) is exposed. A little more scraping may be done till the oil-glands are visible.

Following this, any given area of the peel, say 1 cm square (=1 sq.cm) is measured with a scale and marked suitably. This part is cut out with a sharp scalpel, or using a suitably moulded knife. The piece should be placed over a flat surface. Then the number of oil-glands on the under-surface of the piece should be seen through a hand-lens and counted with care. The number of large and small glands countable on unit area of the peel piece may be recorded. Very small glands are however, difficult to be detected and counted.

3.13 Colour of Edible Part

In many fruits, the colour of the edible parts, *i.e.*, juice, pulp or the hard kernel, respectively in case of juicy fruits, pulpy fruits and nuts is considered to be a pertinent character in analytical studies for a number of reasons. The important ones are, firstly the consumers' acceptance is determined much by this trait. Examples may be cited as pink-coloured juice of some grapefruit and shaddock varieties, rich yellow juice of the mandarin orange variety, *Nagpur Santra*, orange-red pulp of the papaya variety, *Co-3*, red-coloured juice of the sweet orange variety *Malta Blood Red*, deep-red pulp of water-melon varieties, pink-coloured pulp of the *Hafsi* and some other varieties of guava, pink or red-colour juice of the pomegranate varieties, *Alandi*, *Kabul*, *Kandahari* etc. and so on.

Secondly, in some fruits, the colour of the inside part undergoes change with progress of their development, maturation and ripening. Accordingly, this character is sometimes considered as an index to determine harvest-maturity of them or the degree of their ripening in pre-harvest or post-harvest condition.

Thirdly, in some fruits, notably mango and papaya, change of the pulp colour to orange-red during ripening has its impact on their nutritive value. Especially in mango, the orange colour which appears in the ripe fruits is primarily due to accumulation of the pigment, carotene, which is a precursor of retinol, *i.e.*, vitamin A in the ripe fruits of papaya also, carotene is in abundance among other pigments, notably carica-xanthin.

The colour in a fruit whether in the inside or the outside part of it is done to various combinations of chlorophylls, carotene, xanthophylls, anthocyanin, anthoxanthin, or other pigments in different degrees, some of which are water-soluble unlike others that are fat-soluble. The quantitative estimations of these various types of plant pigments may be done by chemical analysis. However, in a gross manner, the relative intensity of the colour may also be judged by eye-estimation or the juice in a colorimeter as stated below.

Procedure

To determine colour of the pulp or juice of a fruit, it should be cut at first. If the fruit is a pulpy one, the pulp colour may be assessed by the same method as described earlier in determining peel colour (Ref. 3.7).

For juicy fruits such as citruses, pomegranate, *phalsa* etc. the juice should be extracted from the fruits. This may be done following the method as laid down earlier. On extracting the juice, its colour should be judged by the same method (Ref. 3.7).

However, for greater accuracy and especially when comparison of two or more juice samples is required in respect of their colour, determination of the Optical Density (O.D.) value of them may be done in a colorimeter. For this, the juice should be made absolutely free from pulp pieces by centrifugation or suitable filtration. The absorbance of the clear juice obtained should be measured in a colorimeter against distilled water as the blank. The filter to be used will no doubt, vary according to the colour of the juice and for juice samples of different colours, the filters to be used may be as follows:

Purple coloured juice	Filters :	575 - 615 nm
Blue coloured juice		615 - 660 nm
Blue-green coloured juice		660 - 690 nm
Green coloured juice		690 - 740 nm
Yellow coloured juice		390 - 420 nm
Orange coloured juice		420 - 460 nm
Red coloured juice		460 - 560 nm

3.14 Pulp Firmness

Like firmness of the entire fruit firmness of the pulp is also considered to be an important feature in analytical studies of some pulpy fruits. This is particularly given emphasis for such fruits like *bael*, elephant apple and perhaps some nuts in which cases, the pulp though softens during development, maturation or ripening, little or detectable softness is noticed in the outer covering.

Procedure

The procedural details to determine firmness of the pulp would remain the same as described earlier in determination of the firmness of the uncut fruits (Ref. 3.7).

3.15 Central Cavity

In some fruits, *e.g.*, papaya, some citruses, notably in the *Nagpur Santra* variety of mandarin orange (Figure 19) etc., a cavity is formed in the centre which makes the edible matter reduced. With greater the volume of the cavity, not only the amount of edible part is lesser but also the shelf-life of the fruits. The volume of the cavity in these fruits accordingly receives attention and is determined.

Procedure

In a general way, the width and the length of the hollow space are measured with a suitable scale. A better method may be to

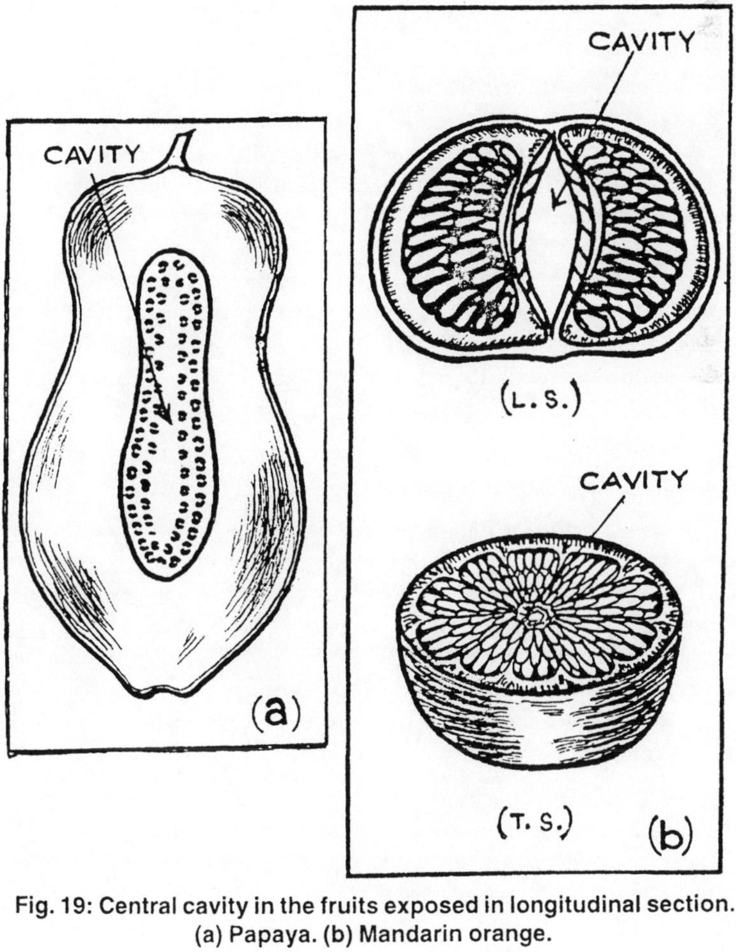

Fig. 19: Central cavity in the fruits exposed in longitudinal section. (a) Papaya. (b) Mandarin orange.

determine the volume of the intact fruit at first without cutting and then again after cutting it, exposing the cavity. On subtracting these two volumes, the volume of the cavity may be known. In determining the volume, the method as described earlier (Ref. 3.2) may be followed.

3.16 Edible Matter Content

Although in fruit culture, the quantitative productivity of a fruit crop is conventionally determined in considering the yield of the whole fruits or the number of them that are harvested from a given number of trees or an unit area of land, it is more apposite to determine how much matter that is consumable to the human beings has actually been obtained from the same trees or the same area of land.

The significance of the above statement lies in the fact that although there are some fruits like guava, seedless grape etc. where the entire fruit has edible value including the peel and the seeds present, in majority of the other fruits, such is not the case and all parts are not consumable to man. Thus, in a mango (Figure 20a), only the mesocarp is consumed while other parts, *i.e.*, endocarp, seed coat, cotyledon or epicarp though may be consumed by animals are not by man. Again, in an apple (Figure 20b) or a pear, the fleshy receptacle, *i.e.*, the thalamus comprises the edible part and the central part containing seeds developed from the carpel is though real pericarp of the fruits remains inedible to man.

Below is given a list of some fruits, the botanical part of which is actually edible to the human beings (Table 1).

To what proportion the human consumable matter is actually present in a fruit is in fact, of practical concern to a fruit scientist dealing with food. This varies greatly according to species, varieties under them, rootstocks used, edapho-climatic conditions of growing, agro-inputs applied and so on. Table 2 has provided in general, the percentage of *edible matter* obtainable from some fruits. The figures are only gross and should not be considered to be applicable under all conditions.

It would be apparent from the Table 2 that *if the amount of edible matter that may be actually available from a fruit is brought for consideration, the productivity as whole fruits will be on the lower side than the productivity as human food for fruits (human consumable part) under most of the species.*

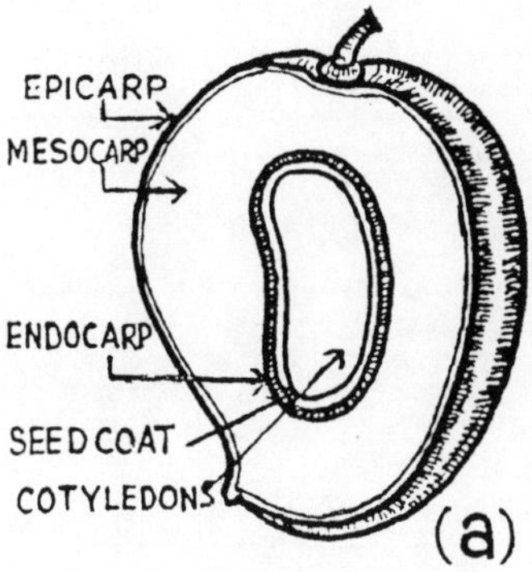

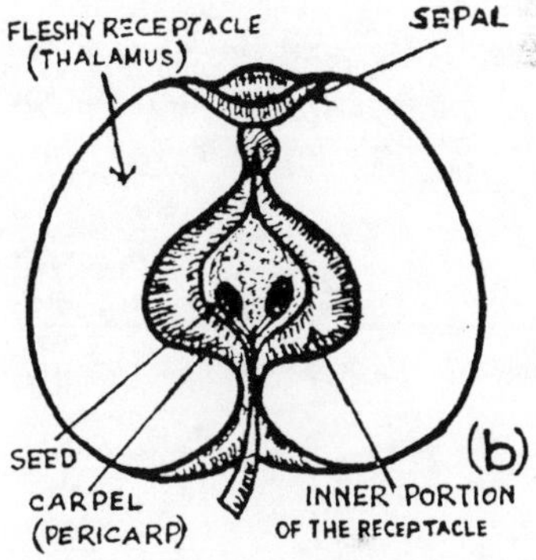

Fig. 20: Fruits parts of (a) mango and (b) apple.

Table 1: Botanical (morphological) parts of some fruits having edible value to the human beings

Name of fruit	*Botanical part edible to man*
Mango	Mesocarp
Banana	Mesocarp and endocarp
Citruses	Juice in hairs arising from inner wall of endocarp
Guava	Pericarp fused with thalamus
Grape	Pericarp and placenta
Pine-apple	Outer rachis, bract, perianth and pericarp – all fused together
Jackfruit	Bract and perianth (also seeds)
Pomegranate	Succulent testa

Contd............

Table 1–Contd......

Name of fruit	*Botanical part edible to man*
Litchi	Fleshy aril (integument)
Bael, Elephant apple	Succulent placenta and inner pericarp
Cashew	Kernel (in nut) and cashew-apple (fleshy, swollen peduncle)
Apple, Pear	Fleshy thalamus and inconspicuous ovary
Strawberry	Fleshy thalamus
Peach, Plum	Epicarp and mesocarp

Table 2: General productivity in terms of whole fruits as well as human consumable parts in them for some fruits

Name of fruit	*Percentage of human consumable matter*	*Yield (tonnes/hectare/year)*	
		As whole fruit	*As human consumable part*
Apple	95.1	10.2	9.7
Banana	65.2	20.5	13.3
Bael	61.3	63.0	38.6
Ber	78.0	15.0	11.7
Cherry	75.0	9.0	6.8
Custard apple	49.6	15.0	7.4
Date (processed)	87.5	14.3	12.5
Guava	99.8	12.3	12.3
Grape (seeded)	97.6	40.0	39.1
Grape (seedless)	100.0	27.0	27.0
Jackfruit	30.6	201.5	61.7
Jamun	84.6	4.3	3.6
Litchi	59.3	12.0	7.1
Mango	71.0	12.0	8.5
Mandarin orange	52.0 (juice)	9.2	4.8
Papaya	73.6	62.0	45.6
Peach	84.5	8.0	6.8
Pear	79.6	34.0	27.1
Pine-apple	68.0	16.8	11.4
Pomegranate	64.0 (juice)	12.8	8.2

Contd............

Table 2–Contd......

Name of fruit	Percentage of human consumable matter	Yield (tonnes/hectare/year)	
		As whole fruit	As human consumable part
Sapota	84.7	24.0	20.3
Sour lime	57.6 (juice)	58.0	33.4
Sweet orange	54.8 (juice)	14.0	7.7

N.B. Date given above are only on general basis and not to be considered applicable under all conditions

Thus, it may be noted that to a fruit scientist, the *yield as human consumable matter of a fruit crop per tree, or per unit area of plantation* is of actual concern than the total yield of fruits.

Procedure

The sample of fruits under study should at first be weighed. In case of pulpy fruits or nuts, these are cut out and the parts in them, which are not consumable to the human beings, are separated out from those part or parts that are consumable. If any consumable matter remains adhered to the separated non-consumable part, it should be brought back to the former part by careful scraping. Then the consumable part is weighed. The percentage of consumable matter may be worked out as follows.

$$\frac{\text{Weight of consumable matter (g)}}{\text{Weight of whole (intact) fruits (g)}} \times 100$$

In case of juicy fruits, *i.e.*, such fruits where only the juice is consumable and not other parts, e.g. citruses, pomegranate etc. the juice should be extracted from the fruits as described earlier, after the fruits have been weighed. Then the juice sample should be strained through cheese-cloth to get free from seeds, rags etc. (Some pulp-pieces will however, pass through the cloth while straining the juice).

It is pointed out in this context that many fruit scientists are of the opinion that since little amount of fine pieces of the pulp is consumable by human beings along with the juice, especially when squash is made, that which will be present in the juice after straining should be left as such, without getting it removed. In other words, the pulp-pieces present in the juice after straining comprise as the edible matter. With consideration to this viewpoint, the volume of juice including some amount of fine pieces of pulp is measured in a measuring cylinder and its percentage in the fruits is determined as follows.

$$\frac{\text{Volume of juice (ml)}}{\text{Weight of whole (intact) fruit (g)}} \times 100$$

3.17 Juice Content

The amount of juice present in a fruit is considered to be one of the most important qualities in juicy fruits like citruses, pomegranate, *phalsa* etc., where only the juice comprises the human consumable part in the fruit. Besides, in some pulpy fruits also which contain abundant juice, *e.g.*, grape, *jamun*, litchi, strawberry, star-apple, water-melon, cape-gooseberry etc. the juice content in the fruit is given emphasis.

Whether in citruses or in other fruits, the amount of juice varies on a number of factors. These may be due to varieties, edapho-climatic conditions of growing, type of rootstock used, nutritional status of the tree, season of fruit development, agro-inputs applied, attack by pests and diseases and the time, frequency, amount and method of irrigation. In some citruses, delayed harvest of the fruits is also a cause for *granulation of the juice*, resulting lowered juice content in them as is found in some varieties of grapefruit, mandarin or sweet orange and notably in the *Kinnow* variety of *tangor*. The stage of development and maturation of the fruits has also its impact on their juice content.

The amount of juice present in the fruits is expressed as percentage basis and is determined as follows.

Procedure

For juicy fruits

In case of juicy fruits like citruses, pomegranate etc., the juice should at first be extracted from the fruits following the method as described earlier. Then the juice should be strained through cheese-cloth to make free from seeds, rags etc. Some amount of fine pieces of pulp will however, pass through the cloth-strainer.

In determining the edible matter content as stated earlier (Ref. 3.15), the fine pieces of pulp, which pass through the cloth-strainer are considered together with the juice. However, where only the juice content of the fruits is to be determined, it is logical to use only clear juice, making it free from pulp-pieces. In accordance with, the strained juice should be centrifuged or suitably filtered to wipe out completely the fine pieces. Thereafter, the volume of the clear supernatant or filtrate as the case may be is measured in a measuring cylinder. The percentage of juice in the fruits may be worked out as follows.

$$\frac{\text{Volume of juice (ml)}}{\text{Weight of the whole (intact) fruits (g)}} \times 100$$

For pulpy fruits

The fruits should at first be weighed and after cutting them, the non-consumable parts should be separated out. Then the juice from the consumable part is squeezed out with the help of a suitable pressing instrument. The pressing should however, be very thorough, ensuring complete recovery of the juice from the pulp pieces.

The squeezed out juice may be centrifuged or suitably filtered, if necessary and the volume of the supernatant or the filtrate be measured in a measuring cylinder. The percentage of juice in the fruits may be determined, considering the weight of the fruits and the volume of the juice obtained from them.

The above method though simple, may have practical difficulty in the act of squeezing out the juice completely from the fruits. In such case, the following method seems to be useful.

After weighing the fruit, the consumable matter from them should be completely taken out. Then known volume of distilled water is added to that and ground thoroughly in a suitable instrument, ensuring that complete grinding of the pulp has been accomplished. The ground material in water is then centrifuged or filtered through glass-wool or suitable medium and the pulp-pieces are separated out.

The volume of the supernatant or the filtrate, as the case may be is then measured in a measuring cylinder. The percentage of juice in the given sample of fruits may be worked out, as follows:

$$\frac{\text{[Volume of the filtrate (ml)] – [Vol. of distilled water used (ml)]}}{\text{Weight of fruit (g)}} \times 100$$

3.18 Flavour

Flavour is charm to a fruit. For this attribute, even a qualitatively less important fruit may attract consumers. It is however, an interesting fact that liking or disliking of smell in fruits may vary according to persons. An example may be cited with fruits of the famous mango variety, viz., *Langra*. Despite being highly delicious in taste, a section of consumers have aversion to these fruits for the turpentine smell they get from them, which on the other hand, is highly liked by some. Many good varieties of papaya fruits fail to appeal consumers for the characteristic papaya-odour, which although is a trifling matter to others. Same is the case with jackfruit, some varieties of sapota and even those grape berries having so-called *foxy* smell. The smell of the ester, amyl acetate in banana, or the sweet smell of other fruits due to the presence of some alcohols, esters and especially, ethylene is even known to cause nausea or other physical discomfort to some, although such sensitive persons are not largely met with.

Chemically, flavour of a fruit is due to the existence of a diverse type of volatile compounds, *e.g.*, alcohols, esters, aldehydes, ketones, ethers, halides, hydrocarbons and others in different proportions, depending on the species and variety.Though the number of these compounds is usually large, amount is low to very low but even the very small amount may have a marked effect. These compounds are also known to change with development, maturation and ripening of the fruits. The detection of the volatile compounds and their estimation are done by ultra-sophisticated chemical procedure particularly as some of them are in fact, present in extremely small amount. However, for practical purpose, the flavour of a fruit may be detected taking advantage of the smelling power of expert persons.

Procedure

Only those persons who have perfect smelling power and have no addiction to smoking or any type of tobacco, narcotic or drinking should be chosen to act as tester. The testers should be adults and comprise both men and women. Several groups of testers should be formed, each group having at least two members and they should be assigned to smell the fruits and express their view freely in writing.

The reports put forward by the different groups should be collected together and the average of these should be considered.

3.19 Seed Content

Presence of seeds in a fruit is considered a demerit or a merit. It is a well-known fact that the consumers have always a preference for seedless fruits. This has motivated either the fruit-breeders and biotechnologists to create such type of fruits by making alteration in the chromosome content of them, or the fruit-physiologists in arranging supply of the necessary stimulus to the ovaries from exogenous source, disallowing the male gametes to fertilize the eggs. Even mother nature also desired to produce parthenocarpic bananas and some other seedless fruits by enriching their ovaries with the stimulus, thus enabling them to grow without fertilization of sexual units.

On the other hand, seeds are valued at least in some fruits for certain purposes. It is noted that seedlessness may lessen the size or qualitative value of some fruits and notably, in custard apple, jackfruit, strawberry and perhaps papaya also. Experimental evidence provides proof to the fact that the seedless berries of custard apple and jackfruit tend to be less sweet than the seeded berries in the same fruit. Also that seedless guavas are inferior than the seeded fruits.

Production of viable and bold seeds in adequate numbers is however, indispensably required for papaya and some other fruit crops which are conventionally propagated by the use of seeds. To produce seedling plants for use as rootstocks, seeds are required. Above all, in case of nuts viz., walnut, almond, pecan-nut, pista-nut, hazel-nut, cashew nut, water-chestnut, sterculia-nut etc., it is the kernel of the seed that comprises the human consumable part. Example of jackfruit may be cited in this connection, which though is not a nut, produces seeds which have edible value to the Indian people. Besides, seeds of apricot yield edible oil and those of custard apple are used for extraction of oil that has many industrial uses. The importance of seeds has been there in hybridization and breeding procedure as well.

Procedure

The conventional way is to take the seeds after cutting the fruits, barring in strawberry, where achenes may be obtained without cutting. If any pulp is adhered to the seeds, it should be removed by rubbing with wood-ash and washing. After washing, the water should be wiped out from their surfaces with a blotting or tissue paper. Records to keep usually include the number of bold and less bold seeds per fruit and the weight of one hundred seeds. Viability of the seeds is also sometimes considered.

3.20 Acceptance to Consumers

The acceptability of a given sample of fruits to the consumers has great relevance for they comprise food to them. In many analytical studies therefore, the fruit scientists lay emphasis on whether and to what extent the materials under any treatment make their appeal to the consumers in respect of quality. Conventionally, the extent of consumers' liking is based on *overall performance* of the fruit only in its *intent* condition. That is, the quality of the fruit materials are to be judged by them, by observing, touching and smelling, without damaging or tasting them, in as much as such is the practice that is mostly followed in a market.

It however, needs mentioning in this connection that the consumers in the market constitute a diverse group of individuals. The diversity of people exists in fact, on a multiplicity of factors, some of which may be stated as age, gender, family condition, community, locality of living, occupation, socio-economic status, affliction to any disease and so on. It is a common observation that preference to a sample of fruits is much related to these factors. For example, a child or a woman may have greater aptitude to a rich-coloured fruit than what a man has. Again, flavour of the fruit is an important attribute for its liking by the fair sex, large-sized fruits by men and in a fair way, the relatively soft-textured fruits by the elderly people. Therefore, in a fruit analytical study undertaken in a given locality, due emphasis to get the consumers cannot be connived at. The general way to make such a test is stated below.

Procedure

The judging which is done in a score-carding manner should evidently got to be carried out by testers who are the actual consumers in the market. The consumers in the market may in fact, be categorized under a large number of classes, in respect of the criteria considered for such groupings and it is a logical approach to get the tests done by individuals belonged to as many classes as possible. It should also be noted that greater the resemblance among the testers is in a class, fairer the result is expected to be.

However, in case of practical difficulty to constitute so many classes of testers in a market, formation of three classes may be resorted to depending on the socio-economic condition of the consumers. These may be graded as of high income, medium income and low income groups. Each of these classes may further be sub-divided as those consisting of children, adolescents, adults and elderly people falling under both genders. Thus, each of the three classes may have four groups of male as well as four groups of female members, making a total of twenty four groups.

Again, each group should have several sub-groups in it as replications of the study and at least two testers should be included in a sub-group. The testers should be assigned to judge the overall performance of the given sample of fruits as is conventionally followed in the market, *i.e.*, by the act of touching, feeling with light pressure and smelling without getting them tested or damaged.* The observations should be recorded by them as desired, however keeping in view the instruction, if any has been given to them by analysts.

The observational reports should then be obtained from the testers under each sub-group belonged to a group and these should be compiled together to determine the average viewpoint put forward by a class of consumers. It is again advisable to get this task carried out by a panel of persons who are known to be expert in reviewing.

* Testing of fruits may be permitted, if *organo-leptic* test of them is desired.

4

Making Representative Sample of Fruit Tissue for Chemical Analysis

In fruit science (pomology), the various types of fruits that are consumable to the human beings are conventionally classified into the following three groups, in consideration to the parts of them that are consumed. (The preserved fruits however, are not considered into these groups).

1. Pulpy fruits
2. Nuts
3. Juicy fruits

Pulpy fruits

The pulpy fruits consist of those, fleshy pulp of which serves as the edible part. Thus, the vast majority of the fruits, *e.g.*, mango, banana, papaya, guava, pine-apple, jackfruit, grape, litchi, custard apple, apple, pear, peach, plum etc. come all under this group. Whether the edible part is fleshy or not has been the only criterion for this group, regardless of the botanical part. Thus, *cashew-apple* (Figure 21), which is botanically the fleshy swollen peduncle and never recognized as a fruit has also reason to come under this group from pomological standpoint.

Nuts

The nut is recognized in Botany as a one-chambered, one-seeded fruit that is developed from a polycarpellary ovary and has a hard and wooly pericarp. In pomology also, the same condition is almost accepted and the *kernel* present inside the nut forms as the edible part. Examples of some common nuts are hazel-nut, walnut, cashew-nut, almond, pecan-nut, chestnut, pista-nut etc. However,

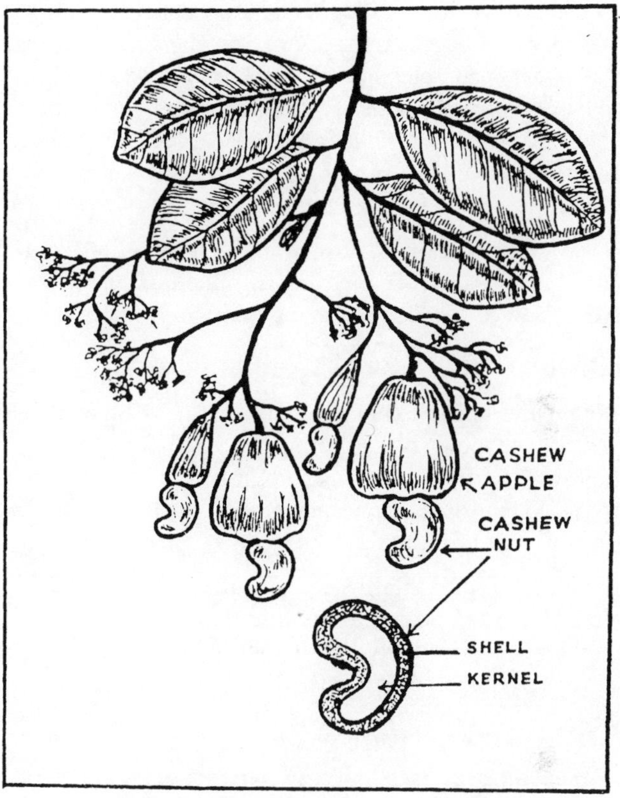

Fig. 21: Cashew-apple, nut and kernel

some botanists consider litchi* as a nut, but it is not consi lered so in pomology, where it is grouped as a fleshy or pulpy frui .

Juicy fruits

The juicy fruits comprise those, in which case, only the juice of the fruits forms the consumable part. Few examples are there under this group, among which, the citrus fruits are best known. Apart from citruses, in pomegranate also, it is only the juice that is consumed. Only the juice is taken in *phalsa* and perhaps in few other minor fruits also.

---

* In China, there is a practice to dry intact litchis. When dried, the edible part, known as *aril*, becomes brown and leathery and the skin hard. This is popularly known as litchi-nut but this is not a real nut like other examples

A botanist or a plant biochemist may consider any part of a fruit for analysis. However, this is not done by a fruit scientist. *For the purpose of conducting any analytical study with a fruit, the fruit scientist uses that part of it which is actually consumed by the human beings. Thus, in a fleshy or pulpy fruit, the pulp, in a nut, the kernel and in a juicy fruit, the juice forms the part for analysis and the results are also expressed in terms of pulp, kernel and juice respectively.*

Now, the question is, how to make representative samples from these parts of the fruits for carrying out the assays. The method generally followed is described in the following.

Procedure

Regardless of whether the fruit that is to be analysed is a pulpy, a nutty or a juicy one, the first step is to take intact fruits and only those intact fruits are to be used for this purpose, which have been prepared to form as *representative sample*, following the method as described under Chapter 2.

From the representative sample of intact fruits which have been already made, the representative sample for conducting the actual analysis, respectively for each of the three groups of fruits may be done in the following way.

Pulpy fruits and nuts

First of all, the fleshy edible part in case of a pulpy fruit while the edible kernel in case of a nut should be scrupulously taken out from the given sample of fruits. The non-edible parts of them have no use and hence, are discarded.

Then the pulp or the kernel as the case may be should be cut into very small pieces after which, the pieces should be thoroughly mixed together and piled at a place (Figure 22a). The pile is then disrupted to make several lots from it at *random basis* (Figure 22b). From each lot, some pieces are picked up randomly and these are mixed together and heaped again (Figure 22c) while the remaining ones are discarded. This heap of pieces is broken and several lots are made in the same way as before. This process is repeated several times and lastly, the *representative sample* for analysis is obtained. Thus, the process follows the similar line of action as has been done in making representative sample of intact fruits described earlier (Ref. Chapter 2).

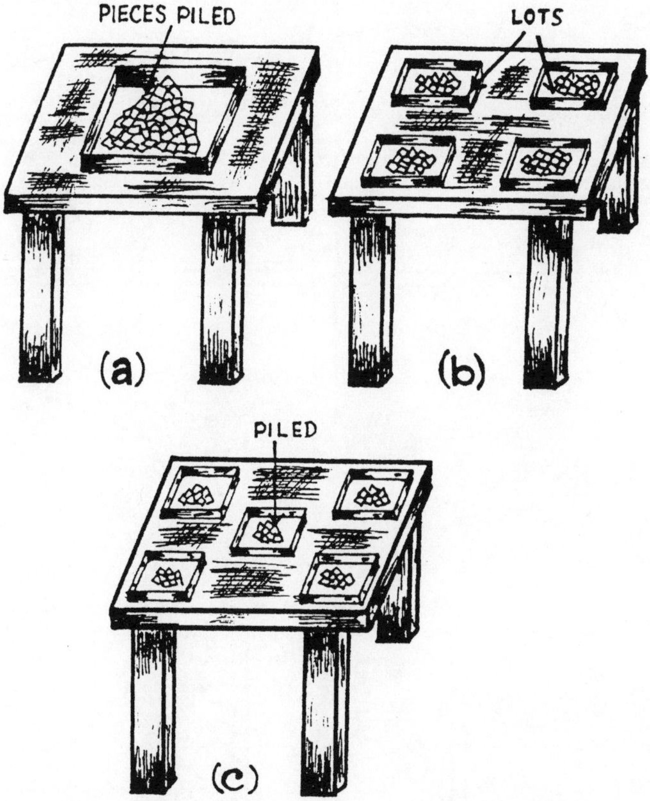

Fig. 22: Making representative smaple of fruit tissues.
(a) Pieces made from intact fruits are mixed and piled together. (b) Pile is disrupted to make several lots. (c) Some pieces taken from each lot are mixed and piled again.

Juicy fruits

In case of juicy fruits, extraction of juice is needed to be done at first. This should be done in the following manner.

From citrus fruits

For a citrus fruit of any species or variety, a number of types of juice extractors are available and the one which can be conveniently used may be taken.

In one type, as shown in Figure 23a, the fruit is at first cut across in the centre to make two halves (Figure 23b) and the juice

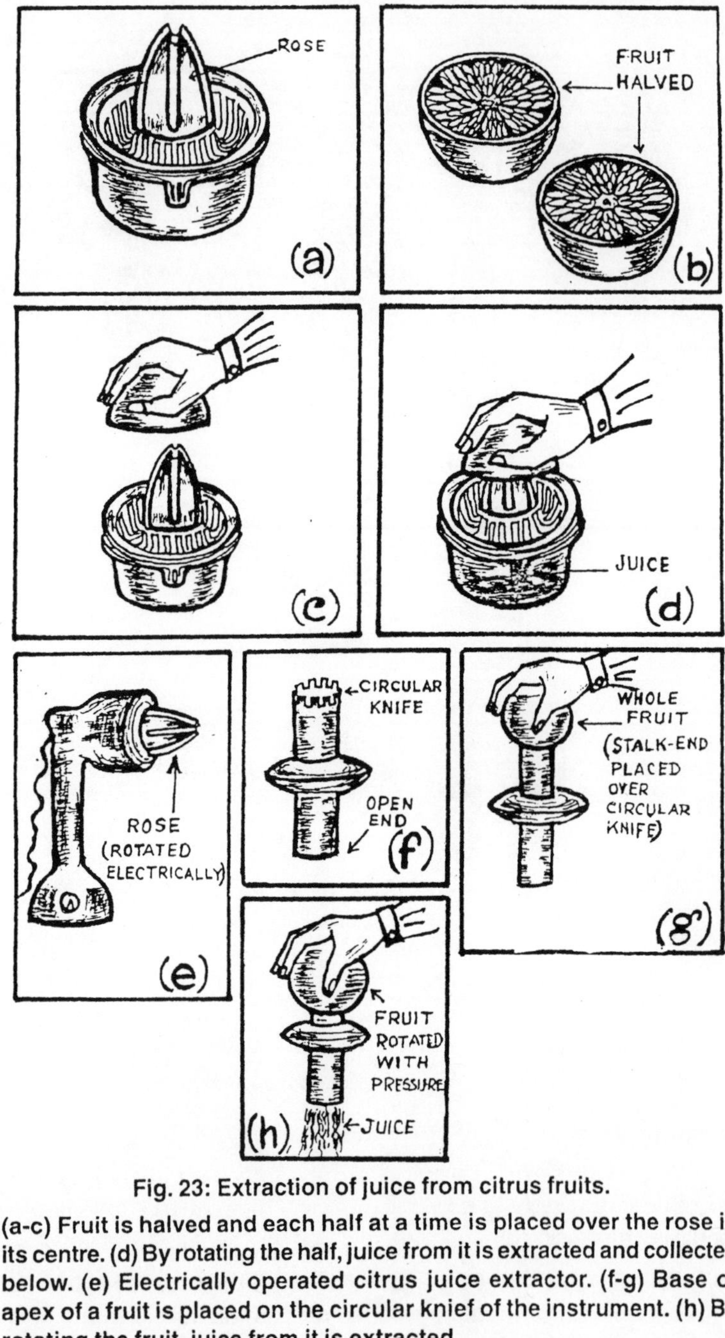

Fig. 23: Extraction of juice from citrus fruits.

(a-c) Fruit is halved and each half at a time is placed over the rose in its centre. (d) By rotating the half, juice from it is extracted and collected below. (e) Electrically operated citrus juice extractor. (f-g) Base or apex of a fruit is placed on the circular knief of the instrument. (h) By rotating the fruit, juice from it is extracted.

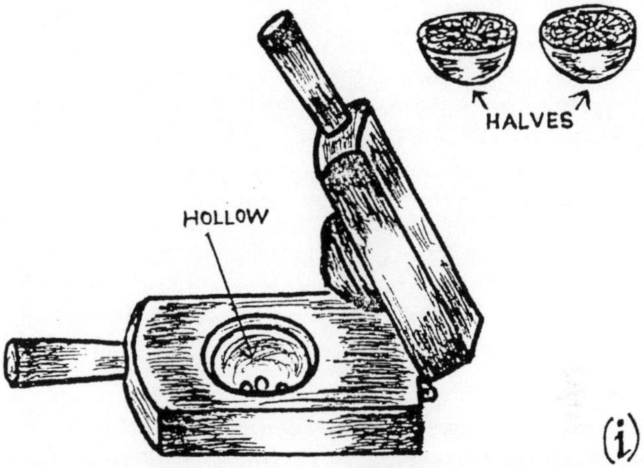

Fig. 23: Extraction of juice from citrus fruits.

(i) Each half at a time is placed in the groove. Then covered with a lid, pressure is given and juice is squeezed out.

from each half is extracted at a time. To do so, the outside of the half is held in palm and the central point of it on the cut surface is brought over the *rose* of the extractor (Figure 23c). Then, by rotating the half around the rose with manual pressure, the juice is easily extracted (Figure 23d) from it. The instrument is available in glass, plastic or stainless steel and although manually operated, recovery of juice is done almost completely. Electrically operated extractor of such type is also available (Figure 23e) having greater efficiency.

There is another type (Figure 23f) which is especially suitable for lime or other small-sized citrus fruits. At one end of the plastic made instrument, there is a circular knife and to work with the knife is pierced at base or apex of the fruit in the centre with one hand and the fruit is held with the other hand (Figure 23g). Then the fruit is rotated around the instrument with manual pressure, due to which the juice is extracted and flows outside through the other opening of it (Figure 23h).

The juice of lime or of a small-sized citrus fruit may also be recovered with a wooden-made press (Figure 23i). The fruit is halved and each half is placed at a time in the groove of it. It is then covered with the lid, manual pressure is given over the handle and the juice from the half is extracted, which comes outside through the lower opening of the instrument.

From pomegranate and other fruits

A commonly available fruit juice extractor (Figure 24a) having perforations at the bottom and the lower part of it is convenient for squeezing out juice from pomegranate, *phalsa* and other fruits. For pomegranate, the berries should be separated out to squeeze their juice while for others, intact fruits may be used.

The berries or the intact fruits as the case may be are to be taken in the cup of the extractor (Figure 24b) to its capacity and it is covered with the lid. Then pressure is given on the handle (Figure 24 c) due to which, the juice will be squeezed and will come outside through the perforations which is collected below.

After squeezing, the lid should be opened and the berries or the fruits should be stirred to turn the upper ones to the lower and the lower ones to the upper layers. Then again it is covered with lid and pressed to squeeze out the juice. The process is repeated several times till the recovery of juice is complete.

Filtration of the juice

The extracted juice need to be made free from pulp-pieces, for which, it should be filtered to get the clear juice having no opacity.

Making representative sample

Some quantity of filtered juice extracted from different fruits belong to the given sample is taken and mixed up. This is stirred to make homogenized and is considered as the representative sample. As the ingredients to be assayed remain in the juice in soluble form, repeated grouping and pooling may not be required as in case of pulpy fruits.

Unitary amount of Analytical Sample

From the above discussion, it is now clear that for the purpose of analysis, the representative sample should be taken either in solid or in liquid form. Evidently, for citrus, pomegranate and few other fruits, the sample should be used in liquid form and for vast majority of all other fruits including nuts, the sample should be taken in solid form.

Now, for quantitative assay of any biochemical component or to determine the activity of any enzyme, it is essential to use a known amount of representative sample.

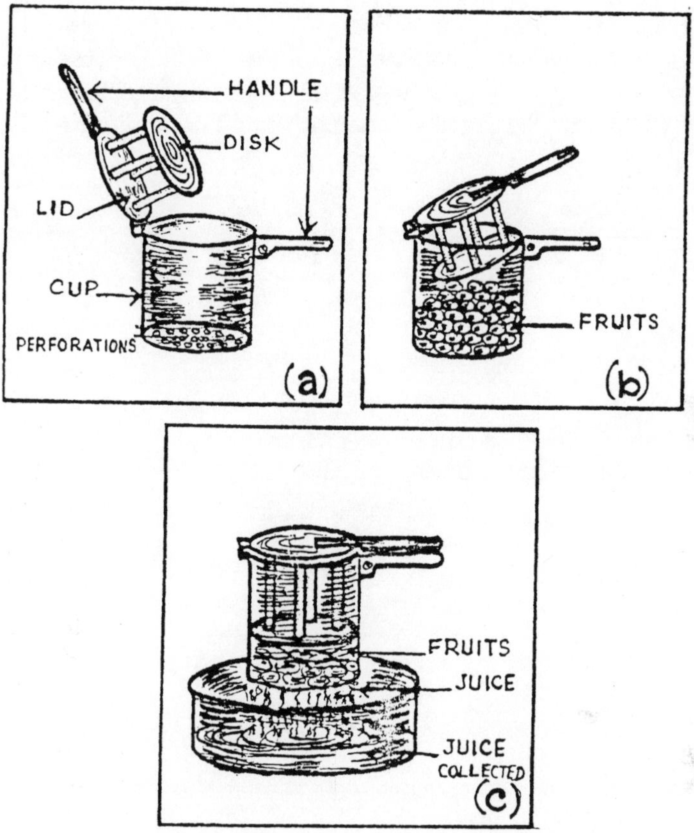

Fig. 24: Extration of juice from pomegranate and other fruits. (a) Extractor. (b) Berries or fruits placed in the cup. (c) Covered with lid, pressure is given on handle and squeezed out juice flows outside through perfortations.

Needless to say, for a solid material, the representative sample should be taken by weighing and for a liquid material, it should be taken by volume. (If any gas is used for analysis, it should also be taken in volume).

In the analytical procedure of the biochemical components or enzymic activity of fruits which have been presented in the subsequent part of the book, the methods have been described considering weight of the representative tissue sample as is done in case of pulpy fruits or nuts. For juicy fruits however, this should be considered in volume. Thus, it should be noted that 1 gram of sample by weight should be considered as 1 millilitre when it is to be taken in volume. (Similarly, for gas also, it should .be considered as 1 millilitre).

5

Preparatory Aspects for Chemical Analysis

5.1 Solution

By the term, solution is meant in chemistry a *homogenous mixture* of two or more substances in such a finely mingied state that they cannot be separated into the constituent parts by easy mechanical means like filtration, decantation, etc. The relative amount of them may however, be gradually changed within certain limits.

A solution is always composed of two types of constituents, which are the *solvent* or the base and the *solute, i.e.,* the substance which is present in the solvent in dissolved state. Both the solvent and the solute may be liquid, gaseous or solid substances. Also that, there may be one or more than one type of solute or solvent in the same solution in dissolved state.

Every solution has a strength which may be known or unknown and when known, it is termed as a *standard solution*. The different types of standard solutions that are used in fruit analytical procedure are mentioned below with salient features of their preparations.

Percentage Solution

The part of the solute given in any unit, which is present in 100 parts of a solution of the same or equivalent unit is designated as percentage solution, meaning thereby, the *percentage of solute* that is present in the solution. Thus, in metric system, if the weight is considered in gram (g) and the volume is considered in millilitre (ml), a given percentage solution refers to the amount of solute in gram or millilitre, that is present in 100 ml or 100 g solution of the solute. Some examples are given below.

Example 1: Preparation of a 5% solution of sodium chloride in distilled water (Weight/Volume).

[The solute, sodium chloride (NaCl) is solid and hence, it is taken by weight. On the other hand, distilled water is liquid and hence, it is taken by volume. The unit of weight has been considered here in gram (g) while the unit of volume as millilitre (ml)].

Thus, to make a 5% solution of sodium chloride, 5 g of sodium chloride is weighed. It is taken in a beaker of say, 250 ml capacity. About 60 ml of distilled water is slowly poured into the beaker. It is stirred well with a glass-rod till the solute dissolves completely in solvent, leaving no residue. Then it is transferred to a volumetric flask of 100 ml capacity. The volume is made upto 100 ml, *i.e.*, upto the mark with distilled water, washing the breaker repeatedly with distilled water. The solution is thoroughly homogenized. It is now a 5% solution of sodium chloride in distilled water.

(Since the solute, sodium chloride has been taken in weight and the solvent, distilled water has been taken in volume, the solution is represented as weight/volume, *i.e.*, W/V).

Example 2: Preparation of a 5% solution of ethanol in distilled water (Volume/Volume).

[The solute, ethanol (C_2H_5OH) is liquid, for which it is taken by volume and the distilled water being liquid is also taken by volume. The unit of volume for solute and solvent here is considered in millilitre (ml)].

Thus, 5 ml of ethanol is measured by volume. It is taken in a beaker of say, 250 ml capacity. About 60 ml of distilled water is slowly poured into it. On stirring with a glass-rod, it is transferred to a 100 ml volumetric flask. The volume is made upto the mark, *i.e.*, 100 ml with distilled water, repeatedly washing the beaker with distilled water. The solution is thoroughly homogenized. Now, this becomes a 5% solution of ethanol in distilled water.

(Since the solute, ethanol has been taken in volume and the solvent, distilled water has also been taken in volume, the solution is represented as volume/volume, *i.e.*, V/V).

Molar Solution

Molecular weight of a chemical is the weight, which expresses how many times a molecule of it is heavier than one atom of hydrogen. This is the sum of atomic weights of the elements which are components of it.

Therefore, molecular weight =

$$\frac{\text{Weight of 1 molecule of the chemical}}{\text{Weight of 1 molecule of hydrogen}}$$

When molecular weight is expressed in gram, it is called, *gram-molecular weight*, or, *gram-molecule* (also as gram-mol). For example, molecular weight of NaCl = 23 + 35.5 = 58.5. So, gram-molecule of NaCl is 58.5.

Now, *molar* solution of a chemical refers to the number of gram-molecules of it present in 1 litre, *i.e.*, 1000 ml solution of that chemical. Thus, if 1 gram-molecule of the chemical (as solute) is present in 1 litre solution, it is a 1 *molar* solution. Similarly, if 2 gram-molecules are present in 1 litre solution, it is a 2 *molar* solution, if 0.5 gram-molecule is present in 1 litre solution, it comes to a 0.5 *molar* solution and so on.

Example: Preparation of a 1 *molar* solution of sodium chloride in distilled water (Weight/Volume).

(Molecular weight of the solute, NaCl is 58.5 and gram-molecular weight of that 58.5 g).

Thus, 58.5 g of NaCl is weighed and taken in a beaker of 1 litre capacity. Distilled water is added to the beaker slowly to about 600 ml and stirred well with a glass-rod till the chemical has completely gone into solution. The solution is then transferred to a 1 litre capacity volumetric flask. The volume is made upto the mark, *i.e.*, upto 1 litre with distilled water, washing the beaker repeatedly with distilled water. On homogenization, it becomes a 1 *molar* solution of NaCl in distilled water (W/V).

Molal Solution

Such a solution is rarely used in fruit analytical procedure. In *molal* solution, weight of both the solute and the solvent are considered. Both are separately weighed and mixed to make the solution.

Example: Preparation of 1 *molal* solution of sodium chloride in distilled water (Weight/Weight).

Here, 58.5 g of NaCl is weighed and taken in a beaker of 2 litres capacity. 1000 g of distilled water is added to this and stirred well with a glass-rod till the chemical has completely gone into solution. It is homogenized. It becomes a 1 *molal* solution of NaCl in distilled water (W/W). (A *molal* solution of any strength should be expressed stating altitude and latitude of the place where it is made).

Normal Solution

When equivalent weight of a chemical is expressed in gram, it is referred to as *gram-equivalent* and a normal solution of a chemical refers to the number of *gram-equivalent* of it present in 1 litre solution of that. Thus, in a 1 normal solution of a chemical, 1 gram-equivalent, in a 2 normal solution, 2 gram-equivalent and similarly in a 0.5 normal solution, 0.5 gram-equivalent and in a 0.1 normal solution, 0.1 gram-equivalent of the chemical is actually present as the solute in 1 litre solution of that.

Example 1: Preparation of a solution of sulphuric acid of known normality with distilled Water (Weight/Volume).

Here, we should note that equivalent weight of an acid is the number of parts by weight of the acid, which contains 1.008 (say, 1) part by weight of replaceable hydrogen. Again, the basicity of an acid is the number of replaceable hydrogen atoms that the molecules of the acid contains.

Therefore, equivalent weight of an acid may be stated as

$$\frac{\text{Molecular weight of the acid}}{\text{Basicity of the acid}}$$

In case of sulphuric acid (H_2SO_4), molecular weight is 98 and basicity is 2. Hence, equivalent weight of it is 98/2 = 49.

Now, if we are to make 1 normal solution of sulphuric acid, 49 g of H_2SO_4 should be taken. It should be dissolved in distilled water and the volume should be made upto 1 litre with distilled water to make 1 *normal* solution of it. But as a matter of fact, H_2SO_4 is available in liquid form and usually, 100 ml of this acid contains 184 g of H_2SO_4. Accordingly, to take 49 g of H_2SO_4, we have to take 26.63 ml volume of this acid (= 100 ml × 49 g ÷ 184 g). This volume of sulphuric acid (*i.e.*, 26.63 ml) should be taken and the volume of this should be made upto 1 litre by addition of distilled water in a 1 litre volumetric flask. This is homogenized to get a 1 normal solution of sulphuric acid (H_2SO_4) in distilled water.

However, the pure form of the concentrated sulphuric acid (H_2SO_4) which is available in the market (for use in laboratory works) is known to have a normality of 36 (or very close to 36). Therefore, to make a 1 normal solution of sulphuric acid, 1 ml of this should be used and 35 ml of distilled water should be added to this. This is homogenized to get a 1 normal solution of sulphuric acid. In the same way, to make a 2 normal solution, 2 ml of the acid should be taken and 34 ml of distilled water should be added and to make a 0.5 normal, 0.5 ml of the acid should be taken and 35.5 ml of distilled water should be added to this and homogenized.

Example 2: Preparation of a solution of sodium hydroxide of known normality with distilled water (Weight/Volume).

Equivalent weight of an alkali is the number of parts by weight of it, which neutralizes completely 1 equivalent weight of an acid. The basicity of the alkali is the number of equivalent of an acid, which can just be neutralized by 1 molecule of the alkali. In other words, the acidity of an alkali is the number of replaceable hydroxyl groups contained in single molecule of it.

Therefore, equivalent weight of an alkali may be stated as

$$\frac{\text{Molecular weight of the alkali}}{\text{Acidity of the alkali}}$$

In case of sodium hydroxide (NaOH), molecular weight is 40 and acidity is 1. Hence, equivalent weight of it is 40/1 = 40.

Now, if we are to make 1 normal solution of NaOH, we should take 40 g of this chemical (available in the form of hygroscopic pellets), dissolve it with about 600 ml of distilled water in a 1 litre beaker by stirring carefully with a glass-rod. When dissolved, this solution should be transferred to a 1 litre capacity volumetric flask and the volume should be made upto the mark with distilled water, washing the beaker repeatedly with it. The solution on making homogenized will be a 1 normal solution of NaOH.

The above method is theoretically all right but has practical limitation. This is because, the NaOH pellets being highly hygroscopic may absorb moisture during weighing under ordinary

---

N.B: In order to make dilution, concentrated acid should be very slowly and carefully poured in water slantwise and not the water in acid which would result in bouncing.

condition and hence, correct weight is difficult to be obtained. Reaction of the pellets with carbon dioxide in the atmosphere is another problem and it is also difficult to get very pure form of the pellets in the market.

To overcome these difficulties, it is advisable to prepare at first a solution of NaOH in distilled water of unknown strength and then diluting it to bring down to the normality as desired. To make this solution of unknown strength to any known normality, titration of the solution with a *primary standard* is necessary, in order to know the amount of distilled water that should be added for dilution. The primary standard solution should be made of an acid which is available in pure and anhydrous form, *e.g.*, oxalic acid, citric acid etc.

The procedure is outlined below. At first, a solution of NaOH pellets in distilled water of unknown concentration is made. This is stored for few days in a tightly-stoppered vial. After few days of storing, it is taken and without disturbing, the upper clear layer which is free from precipitate is slowly and carefully decanted off. This decanted solution should be used to make a solution of known normality. To do so, we must know the actual normality that has been attained by the NaOH solution and this would be known by titrating it with a given volume of a primary standard solution.

If we use citric acid ($C_6H_8O_7$) as the primary standard, we know that its molecular weight is 192 and hence, equivalent weight is (192/3=) 64. Now, a freshly prepared 6.4% solution of citric acid in distilled water (W/V) is made. A 10 ml part of this solution is pipetted and taken in a conical flask. (Evidently, this 10 ml contains 0.64 g of citric acid). Few drops of phenolphthalein indicator are added in the conical flask. The NaOH solution is taken in a washed and rinsed burette. Titration is done to get the end-point as pink colour and its persistence for some time. On repeating the titration, the confirmed burette reading is noted. From the burette reading, the normality of the NaOH solution may be known as follows.

Say, burette reading is 4.8. Therefore, the normality is (10/4.8 =) 2.08 normal. Similarly, if the reading is 13.2, the normality is (10/13.2=) 0.75.

Now, when the normality of the NaOH solution is known, the solutions of different normalities may be made from it by dilution. For example, if the normality is 4.8, to make it to 1 *normal* solution, for every 1 part by volume of the solution (4.8–1.0 =), 3.8 parts of distilled water should be added and similarly, if the 1 solution is 0.75 *normal*, for every 1 part by volume of it, 6 parts by volume of distilled water should be added to make it a 0.1 *normal* (N/10) solution of NaOH.

Example 3: Preparation of a 1 *normal* solution of a salt, sodium carbonate in distilled water (Weight/Volume).

Equivalent weight of a salt is the part by weight of the compound which contains 1 part by weight of replaceable hydrogen or hydrogen equivalent of an active element present in it, *i.e.*, the part by weight of it which contains 1 equivalent of the constituent metal, or the radical equivalent of the metal.

Therefore, equivalent weight of the salt =

$$\frac{\text{Molecular weight of the salt}}{\text{Total number of basic atoms} \times \text{Valency of the basic part}}$$

In case of sodium carbonate (Na_2CO_3), molecular weight is 106, number of basic atoms is 2 and the valency is 1. Hence, equivalent weight of it is $106/2 \times 1 = 53$.

To make 1 *normal* solution of Na_2CO_3, 53 g of the salt is weighed and taken in a beaker. Little amount of distilled water is added to this, shaken well with a glass-rod till it dissolves completely. The solution is transferred to a 1 litre volumetric flask and the volume is made upto the mark, washing the beaker repeatedly with distilled water. It is then homogenized.

Part Per Million Solution

Part per million (abbreviation: p.p.m.) solution of a solute refers to the part by weight (if it is solid), or part by volume (if it is liquid or gas) of it present in 1 *million i.e.*, 10 lakh parts by weight of the solution (if it is solid), or volume of the substance (if it is liquid or gas).

Example 1: Preparation of a 100 ppm solution of sodium chloride in distilled water (Weight/Volume).

[The solute, sodium chloride (NaCl) being solid is taken by weight, while distilled water being liquid is taken by volume. The unit of weight is considered in gram (g) while that of volume, by millilitre (ml)].

Thus, 100 g of NaCl is weighed and taken in a 1500 litre capacity container. Distilled water is added to this to about 100 litres and stirred well till the solute has completely come into solution, leaving no residue. The volume of the solution is made upto 1 million, *i.e.,* 10 lakh millilitres), in other words, 1000 litres with distilled water. This is homogenized to get a 100 ppm solution of NaCl in distilled water (W/V).

Therefore, 100 g of NaCl (solute) is present in 1000000 ml solution of it in distilled water.

The above solution is however, of extremely high amount and so much solution may not be required. Hence, to make lesser amount of solution, smaller units should be used.

For example, instead of 100 g, say 100 mg of NaCl is taken by weight. It is transferred to a 1 litre beaker. Distilled water is added to it to about 700 ml and stirred well with a glass-rod till the solute has completely gone into solution.

The solution is transferred to a 1 litre volumetric flask and the volume is made upto the mark with distilled water washing the beaker repeatedly with it. On homogenization, a 100 ppm solution of NaCI in distilled water is obtained.

Thus, it is apparent that if the solute is solid and the solvent is liquid, a ppm solution may be expressed as follows.

Solution of a solute in gram/1000 litres (g/1000 l),
or, Solution of a solute in milligram/litres (mg/l),
or, Solution of a solute in microgram/millilitre (µg/ml).

Example 2: Preparation of a 100 ppm solution of ethanol in distilled water (Volume/Volume).

[The solute, ethanol (C_2H_5OH) as well as the solvent distilled water are liquid and hence, both are taken in volume. The unit of the volume is considered in millilitre (ml)].

If 100 ml of ethanol (solute) is taken, the volume of the solution should be made upto 1000000 ml (*i.e.,* 1000 litres) by adding distilled water.

However, in smaller unit, say, 0.1 ml of ethanol is taken in a 1 litre beaker. Distilled water is added to it to about 700 ml and stirred well. The solution is transferred to a 1 litre volumetric flask and the volume is made upto the mark with distilled water washing the beaker repeatedly with it. On homogenization, a 100 ppm solution of ethanol is obtained.

Thus, if both the solute and the solvent are liquids, a ppm solution may be expressed as follows.

Solution of a solute in millilitre/1000 litres (ml/1000 l),

or, Solution of a solute in microlitre/litres (µl/l),

or, Solution of a solute in milli-microlitre, *i.e.*, nanolitre/millilitre (µl/ml).

Making Dilution of Solution

For many solutions, it is desirable that the stock solutions of relatively high concentrations are stored and when needed, these may be conveniently diluted to the desired strength as follows.

Example 1: Stock solution is 9%. By dilution, 3% solution is to be made from it.

To do so, 3 parts by volume of the 9% stock solution is taken. To this, (9–3 =) 6 parts by volume of solvent is added and this is homogenized to get a 3% solution. That is, for 3 parts by volume of the 9% solution, 6 parts by volume of solvent are added. Hence, for every part of stock solution taken, 2 parts of solvent should be added by volume.

Example 2 : Stock solution is 1 *molar*. By dilution, 0.25 *molar* solution is to be made from it.

In this case, 0.25 part by volume of the stock solution is taken. To this, (1.00 - 0.25 =) 0.75 part by volume of solvent is added and this is homogenized to get a 0.25 *molar* solution. Thus, for every 1 part of the stock solution, 0.3 part of the solvent should be added by volume.

5.2 Indicator

In the process of titration in an acid-base reaction, a point is reached when the reaction has just been completed. This point of titration is called the *end-point* and this is usually confirmed by the use of an *indicator*. An indicator is therefore, a chemical substance which by bringing about a change in colour indicates the end-point, in other words, completion of a chemical reaction.

N.B. It should be noted that the concentrations are changed in solutions of [illegible] solutes in storage. For such solutions, concentrations should first be checked before making dilutions from them.

Some commonly used indicators and their *p*H range have been enlisted below (Table 3).

Table 3: Indicator, colour and *p*H range

Indicator	Colour		*p*H range
	Acid	Base	
Thymol blue	Red	Yellow	1.2 - 2.8
Bromophenol blue	Yellow	Red	3.0 - 4.6
Litmus	Red	Blue	4.5 - 6.5
Phenolphthalein	Colourless	Pink	8.3 - 10.0
Cresol red	Yellow	Red	7.2 - 8.8
Methyl orange	Orange	Yellow	3.1 - 4.4

5.3 Buffer Solution

Solutions of many components can resist the change of *p*H, *i.e.*, hydrogen-ion concentration, if a small amount of acid or alkali is added to them. This resistance to *p*H change is known as *buffer action* and a solution which shows this property is referred to as a buffer solution. Buffer solutions are mostly considered of a mixture of weak acid or alkali and its salt. The salient features of a buffer solution may be stated as below.

1. High level of purity.
2. Readily soluble in water.
3. Enzymically and hydrolytically stable.
4. Adequate buffer-action within the specified *p*H range.
5. Should not absorb light of visible spectrum, or ultra-violet ray.
6. Have no toxicity and inhibiting action in biological process.

Preparation of Some Common Buffers

Phosphate Buffer:

Stock solutions

A. Monobasic sodium phosphate (NaH_2PO_4, H_2O) solution: 0.02 *M* (27.8 g of it is dissolved in distilled water and volume is made upto 1 litre with distilled water).

B. Dibasic sodium phosphate ($Na_2H\ PO_4$) solution: 0.02 *M* (53.65 g of it is dissolved in distilled water and the volume is made upto 1 litre with distilled water).

X ml of A and Y ml of B are mixed together thoroughly and volume is made upto 200 ml with distilled water (Table 4).

Table 4: Preparation of phosphate buffer

X ml	Y ml	pH	X ml	Y ml	pH
93.5	6.5	5.7	45.0	55.0	6.9
92.0	8.0	5.8	39.0	61.0	7.0
90.0	10.0	5.9	33.0	67.0	7.1
87.7	12.3	6.0	28.0	72.0	7.2
85.0	15.0	6.1	23.0	77.0	7.3
81.5	18.5	6.2	19.0	81.0	7.4
77.5	22.5	6.3	16.0	84.0	7.5
73.5	26.5	6.4	13.0	87.0	7.6
68.5	31.5	6.5	10.5	89.5	7.7
62.5	37.5	6.6	8.5	91.5	7.8
56.5	43.5	6.7	7.0	93.0	7.9
51.0	49.0	6.8	5.3	94.7	8.0

Acetate Buffer:

Stock solutions

A. Acetic acid (CH_3COOH) solution: 0.2 *M* (11.55 ml of it is dissolved in distilled water and volume is made upto 1 litre with distilled water).

B. Sodium acetate ($C_2H_3O_2Na, 3H_2O$) solution: 0.2 *M* (27.2 g of it is dissolved in distilled water and volume is made upto 1 litre with distilled water).

X ml of A and Y ml of B are mixed together thoroughly and volume is made upto 100 ml with distilled water (Table 5).

Table 5: Preparation of acetate buffer

X ml	Y ml	pH	X ml	Y ml	pH
46.3	3.7	3.6	20.0	30.0	4.8
44.0	6.0	3.8	14.8	35.2	5.0
41.0	9.0	4.0	10.5	39.5	5.2
36.8	13.2	4.2	8.8	41.2	5.4
30.5	19.5	4.4	4.8	54.2	5.6
25.5	24.5	4.6			

Glycine–HCl Buffer:

Stock solutions

A. Glycine ($C_2H_5NO_2$) solution: 0.2 *M* (15.01 g of it is dissolved in distilled water and volume is made upto 1 litre with distilled water).

B. Hydrochloric acid (HCl) : 0.2 *M*.

X ml of A and Y m1 of B are mixed together thoroughly and volume is made upto 100 m1 with distilled water (Table 6).

Table 6: Preparation of glycine-HCl buffer

X ml	Y ml	pH	X ml	Y ml	pH
25.0	22.0	2.2	25.0	5.7	3.0
25.0	16.2	2.4	25.0	4.1	3.2
25.0	12.1	2.6	25.0	3.2	3.4
25.0	8.4	2.8	25.0	2.5	3.6

Boric Acid – Borax Buffer:

Stock solutions

A. Boric acid (H_3BO_3): 0.2 *M* (12.2 g of it is dissolved in distilled water and volume is made upto 1 litre with distilled water).

B. Borax solution: 0.05 *M* [19.05 g of sodium borate ($Na_2B_4O_7$, $10H_2O$) is dissolved in distilled water and volume is made upto 1 litre with distilled water].

X ml of A and Y ml of B are mixed together thoroughly and volume is made upto 200 ml with distilled water (Table 7).

Table 7: Preparation of boric acid-borax buffer

X ml	Y ml	pH	X ml	Y ml	pH
50.0	2.0	7.6	50.0	22.5	8.7
50.0	3.1	7.8	50.0	30.0	8.8
50.0	4.9	8.0	50.0	42.5	8.9
50.0	7.3	8.2	50.0	59.0	9.0
50.0	11.5	8.4	50.0	83.0	9.1
50.0	17.5	8.6	50.0	115.0	9.2

Tris–HCl Buffer:

Stock Solutions

A. Tris (Hydroxymethyl) amino-methane ($C_4H_{11}NO_3$) solution: 0.2 *M* (24.2 g of it is dissolved in distilled water and volume is made upto 1 litre with distilled water).

B. Hydrochloric acid (HCl) : 0.2 *M*.

X ml of A and Y ml of B are thoroughly mixed and volume is made upto 200 ml with distilled water (Table 8).

Table 8: Preparation of Tris-HCl buffer

X ml	Y ml	pH	X ml	Y ml	pH
50.0	5.0	9.0	50.0	26.8	8.0
50.0	8.1	8.8	50.0	32.5	7.8
50.0	12.2	8.6	50.0	38.4	7.6
50.0	16.5	8.4	50.0	41.4	7.4
50.0	21.0	8.2	50.0	44.2	7.2

5.4 Drying of Analytical Sample

Drying of the fruit tissue may become necessary for conducting some of the analyses. It is also sometimes done for preserving the analytical sample for some period. The conventional method of drying of tissue sample is described below.

Procedure

Drying of a tissue sample is conventionally done in a hot-air oven. The sample is freely exposed to a temperature of 68°–70°C for a period of about 72 hours. Following this, the sample is removed from the oven and transferred straightaway in a desiccator to bring it to room temperature condition.

Determination of dry weight percentage

If the percentage of dry weight of the tissue becomes necessary to be determined, known amount of sample by weight should be used and the weight of it just after drying should be recorded. From the fresh and dry weights of the sample, the percentage of dry weight may be recorded as follows.

$$\text{Percentage of dry weight (over fresh weight)} = \frac{\text{Weight of the sample just after drying}}{\text{Weight of the fresh sample used}} \times 100$$

5.5 Ashing of Analytical Sample

Ashing of the dried fruit tissue may become necessary for quantitative estimation of mineral matter of a sample. The conventional method of ashing of dried tissue sample is described below.

Procedure

Ashing of a dried tissue sample is conveniently done in a muffle-furnace. The dried sample is subjected to a temperature of 500°–600° C for some time. Following this, the sample is removed from the furnace and transferred to a desiccator to bring it to room temperature condition.

Determination of ash weight percentage

If a percentage of ash weight becomes necessary to be determined, known amount of dried sample by weight should be used and it is transferred to a crucible (preferably of silica make). It is then ignited in a muffle-furnace at 500°–600° C for about 4 -6 hours, cooled to room temperature and reweighed.

From the dry and ash weight of the sample, the percentage of ash weight may be recorded as follows.

$$\text{Percentage of ash weight (over dry weight)} = \frac{\text{Weight of ash after ignition}}{\text{Weight of dried sample used}} \times 100$$

5.6 Removal of Pigments

Presence of one or more pigments in a sample of tissue sometimes interfere in chemical analysis and this necessitates their removal before the analytical procedure is taken up.

The chemicals which are specific in wiping out different types of pigments from the samples have been enlisted below.

1. Chlorophyll – Activated charcoal.
2. Carotene – Acetone [80 per cent in distilled water (V/V)].
3. Anthocyanin – Ethanolic HCl [95 per cent ethanol in distilled water (V/V): 1.5 *N* HCl (90: 10) at 4°C in dark].
4. Lycopene – Acetone [80 per cent in distilled water (V/V)].

6

Determination of Chemical Constituents

6.1 Carbohydrate

Carbohydrates are principal compounds in a plant body. Chemically, these are hydrates of carbon and have the general formula as $C(H_2O)n$. They exist in different forms, which are conventionally classified into 3 major groups, viz., monosaccharides, oligosaccharides and polysaccharides. The simplest soluble forms of carbohydrates are the 3 – carbon compounds, glyceraldehyde and dihydroxyacetone. All carbohydrates are polyhydroxy-aldehydes or ketones. Polysaccharides constitute most complex form of carbohydrates of high molecular weight.

Principle

Carbohydrates are at first hydrolyzed (broken down) to liberate simple sugars, which is conveniently done by using dilute mineral acids. After hydrolysis, the simple sugars (hexose) that are obtained are dehydrated to hydroxymethyl furfural. This condenses to form a green coloured solution on reaction with anthrone. The intensity of the colour appeared is then measured colorimetrically to determine the amount of carbohydrate present in the sample.

Reagents

(A) *Hydrochloric acid* (HCl): 2.5 *N*.

(B) *Anthrone reagent:* 0.2 per cent [200 mg anthrone ($C_{14}H_{10}O$) is dissolved in little amount of ice-cold H_2SO_4 and the volume is made upto 100 ml with the solvent].

(C) *Standard glucose solution:* 100 mg glucose ($C_6H_{12}O_6$) is dissolved in little amount of distilled water and the volume is made upto 100 ml with distilled water. From this solution, 10 ml is taken and this is diluted to make 100 ml with distilled water. The concentration of this solution will be 100 µg/ml.

Procedure

1. Known amount of tissue sample by weight is taken. This is hydrolyzed with 2.5 *N* hydrochloric acid (Reagent-A) in a boiling water-bath for 2 -3 hours. Then it is cooled to room temperature.
2. After cooling, sodium carbonate (Na_2CO_3) is added to it in small parts. This is continued till effervescence is ceased.
3. The extract is filtered or centrifuged. The filtrate or the supernatant is collected and made upto a known volume with distilled water.
4. A number of aliquots (0.5–1.0 ml) are taken separately into test-tubes. Each should be made upto 1 ml with distilled water.
5. 4 ml of 0.2 per cent anthrone reagent (Regent-B) is added to each test-tube and then it is heated in a boiling water-bath for 10 minutes.
6. Test-tubes are cooled to room temperature. The green to dark-green colouration will appear on heating the samples. The Optical Density (O.D.) value of the coloured solutions are then measured through 630 nm wavelength in a colorimeter against blank.
7. *Preparation of standard curve:* From the standard glucose solution (100 µg/ml) (Reagent-C), different concentration gradients are made as shown in the following table (Table 9).

To each of these glucose solution, 4 ml of 0.2 per cent anthrone reagent is added. This is heated for 10 minutes in a boiling water-bath to develop the green colouration. The O.D. value of the sample is then measured through 630 nm wavelength in a colorimeter against blank. A standard curve is prepared by plotting the known concentrations of glucose solution against respective O.D. value of each.

Calculation

From the standard curve, the amount of carbohydrate actually present in the sample is determined.

Table 9: Dilutions of glucose solutions to make different concentrations

Conc. gradient (μg/ml)	*Volume of glucose solution taken (ml)*	*Volume of distilled water added (ml)*
0	0.0	1.0
20	0.2	0.8
40	0.4	0.6
60	0.6	0.4
80	0.8	0.2
100	1.0	0.0

6.2 Reducing Sugar

In fruits, both reducing and non-reducing sugars are present in varying amount.

Reducing sugars are those hexose ($6_6H_{12}O_6$) sugars, which can reduce compounds such as alkaline (ammoniacal) silver nitrate solution, cupric salt solution etc. When they make reduction reactions, they themselves are oxidised.

Hexose sugars are divided into 2 main groups, which are aldo-hexose and keto-hexose. Aldo-hexose or aldose contains aldehyde group and keto-hexose or ketose contains ketone group. Aldehydes are strong reducing agents. Hexose sugars which contain aldehyde groups, *e.g.*, glucose, galactose, mannose etc., are reducing sugars. Ketones are however, more resistant to oxidation than aldehydes, because it involves the breaking of a relatively stable C–C bond. Hence, they do not ordinarily reduce alkaline silver nitrate or cupric salt solution. But though fructose contains ketone, it is able to reduce readily as easily oxidizable CO-CH_2OH group is present in it. Hence, fructose reacts as a reducing sugar.

Non-reducing sugar, *e.g.*, sucrose is a disaccharide and cannot reduce alkaline silver nitrate or cupric salt solution

Titrimetric Method

Principle

When sugars are extracted and titrated, the reducing sugars only take part in the reaction in making reduction, but the non-reducing sugars that are present in it, do not take part in reduction reaction and remain as such. Accordingly, only the reducing sugars are estimated by titration.

Reagents

(A) *Ethanol* (C_2H_5OH): 80 per cent in distilled water (W/V).

(B) *Somogyi's reagent:* 56 g anhydrous sodium bi-phosphate (Na_2HPO_4) and 80 g sodium potassium tartrate ($C_4H_4KNaO_6$, $4H_2O$) are taken. To this, 200 ml of 1*N* sodium hydroxide (NaOH) and 500 ml of distilled water are added and shaken well to dissolve. Then, 160 ml of 10 per cent copper sulphate ($CuSO_4$, $5H_2O$) and 360 g anhydrous sodium sulphate (Na_2SO_4) are added to the solution. After this, 200 ml of 0.1 *N* potassium iodate (KIO_3) is added and mixed well. Volume is made upto 2 litres with distilled water. After mixing, the solution is stored in a stoppered bottle for a week and then it is filtered. The filtrate is collected and used in the experiment.

(C) *Lead acetate* [($C_2H_3O_{22}$ Pb, $3H_2O$] *solution:* Saturated in distilled water (W/V).

(D) *Di-sodium hydrogen phosphate* (Na_2HPO_4) *solution:* Saturated in distilled water (W/V).

(E) *Potassium iodide* (KI) *solution:* 2.5 per cent in distilled water (W/V).

(F) *Sulphuric acid* (H_2SO_4): 1.5 *N*.

(G) *Starch* [$(C_6H_{10}O_5)n$] *solution:* 0.5 per cent in distilled water (W/V).

(H) *Sodium thiosulphate* ($Na_2S_2O_3$) *solution:* 0.005 *N* in distilled water (W/V).

Procedure

1. *Extraction of sugars*: Extraction of sugars from the sample is usually carried out using 80 per cent ethanol (Reagent-A). The crushed material is refluxed in a Soxhlet's apparatus (Figure 25) for about 30 minutes. After refluxing, alcohol is wiped out from the extract with the addition of distilled water. The extract is centrifuged. To the supernatant, saturated lead acetate solution (Reagent-C) and saturated di-sodium hydrogen phosphate solution (Reagent-D) are added. After centrifugation, the clear

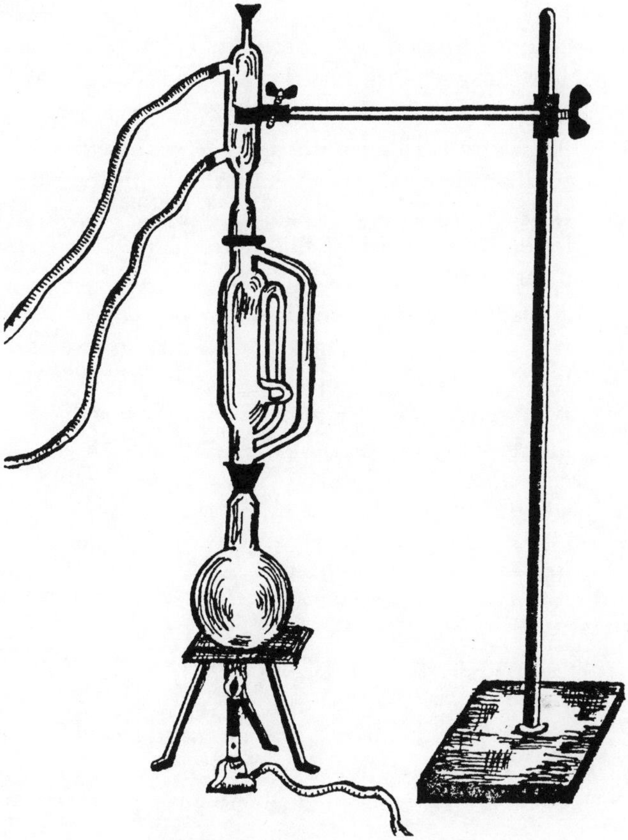

Fig. 25: Soxhlet's apparatus.

supernatant is collected and made upto a known volume with distilled water.

2. To make standard, 1 ml, 3 ml and 5 ml aliquots are taken from the supernatant, respectively in 3 hard-glass test-tubes.
3. 5 ml of Somogyi's reagent (Reagent-B) and distilled water are added so that final volume of the solution in each test-tube comes to 15 ml.
4. The tubes covered with lids are heated in a boiling water-bath for about 15 minutes. These are then cooled to room temperature. The tubes, in which the solutions have red precipitate are discarded.

5. 1 ml of 2.5 per cent potassium iodide solution (Reagent-E) is added to each test-tube. Then 3 ml of 1.5 *N* sulphuric acid (Reagent-F) is added to this and shaken well till the golden yellow colour is formed.
6. A burette is filled with 0.005 *N* sodium thiosulphate solution (Reagent-H). It is titrated with the sample on addition of 1-2 drops of starch solution (Reagent-G) to the latter. The end-point is determined by the complete disappearance of the colour from the solution.
7. A blank is also prepared by mixing 5 ml of Somogyi's reagent (Reagent-B) and 10 ml of distilled water. This is heated in a boiling water-bath for about 15 minutes.
8. After that, it is cooled and 1 ml of 2.5 per cent potassium iodide solution (Reagent-E) and 3 ml of 1.5 *N* sulphuric acid are added to this. Titration is done using 0.005 *N* sodium thiosulphate solution (Reagent-H) containing starch solution (Reagent-G) and the end-point is determined.

Calculation

The amount of sugars present in the sample is calculated by the following formula.

1 ml of 0.005 N sodium thiosulphate solution is equivalent to 0.135 mg, or, 0.000135 g of hexose (reducing sugar).

Therefore, percentage of reducing sugar may be calculated as

$$\frac{0.000135 \times b \times d}{a \times c} \times 100$$

where,

a = Weight of sample.

b = Volume made with distilled water.

c = Volume of aliquot taken for estimation.

d = Non-reacted thiosulphate, *i.e.*, [(burette reading for blank) – (burette reading for sample)].

Colorimetric Methods

Principle

Reducing sugars are at first heated with alkaline copper-tartrate reagent to give rise to cuprous oxide. It is then reacted with arsenomolybdate to form molybdenum blue complex. The intensity of the colour appeared is then measured colorimetrically to determine the amount of reducing sugars present in the sample.

Reagents

(A) *Ethanol* (C_2H_5OH): 80 per cent in distilled water (V/V).

(B) *Arsenomolybdate reagent:* 25 g ammonium molybdate $[(NH_4)_6\ Mo_7O_2,\ 4H_2O]$ is dissolved in 450 ml of distilled water. Then 21 ml of conc. sulphuric acid is added slowly to it and mixed well. 3 g disodium hydrogen arsenate (Na_2HAsO_4, $7H_2O$) is dissolved in 25 ml of distilled water. Solutions of ammonium molybdate and disodium hydrogen aresnate made above are mixed properly and stored at 37°C for 24 hours in dark.

(C) *Lead acetate* $[(C_2H_3O_2)_2\ Pb,\ 3H_20]$ *solution:* Saturated in distilled water (W/V).

(D) *Di-sodium hydrogen phosphate* (Na_2HPO_4) *solution:* Saturated in distilled water.

(E) 2.5 g anhydrous sodium carbonate (Na_2CO_3), 2 g sodium bicarbonate ($NaHCO_3$), 2.5 g sodium potassium tartrate ($C_4H_4KNaO_6$, $4H_2O$) and 20 g anhydrous sodium sulphate (Na_2SO_4) are dissolved in distilled water and the volume is made upto 100 ml with distilled water.

(F) 15 g copper sulphate ($CuSO_4$, $5H_2O$) is dissolved in little amount of distilled water. One drop of conc. sulphuric acid (H_2SO_4) is added to it and the volume is made upto 100 ml with distilled water.

(G) *Alkaline copper tartrate reagent:* The reagent is freshly prepared by mixing 4 ml of Reagent F and 96 ml of Reagent E.

(H) *Standard glucose solution:* 100 mg glucose ($C_6H_{12}O_6$) is dissolved in little amount of distilled water and the volume is made upto 100 ml with distilled water. From this solution, 10 ml is taken and this is diluted to make 100 ml with distilled water. The concentration of this solution will be 100 µg/ml.

Procedure

1. *Extraction of sugars*: Same as in titrimetric method (Ref. 6.2 Procedure → Step 1).

2. A number of aliquots (0.2 → 1.0 ml) are taken separately in test-tubes. Each should be made upto 1 ml with distilled water.
3. 1 ml of alkaline copper tartrate reagent (Reagent-G) is added to each test-tube and then it is heated in a boiling water-bath for 10 minutes.
4. Test-tubes are cooled to room temperature. 1 ml of arsenomolybdate reagent (Reagent-B) is added to each test-tube to develop the blue coloration. The total volume of each test-tube is made upto 10 ml with distilled water. The Optical Density (O.D.) value of the coloured solutions are then measured through 620 nm wavelength in a colorimeter against blank.
5. *Preparation of standard curve*: From the standard glucose solution (100 μg/ml) (Reagent-H), different concentration gradients are made. To each of the solution, blue colouration is developed by the same procedure. The O .D .value of it is then measured through 620 nm wavelength in a colorimeter against blank. A standard curve is prepared by plotting the concentrations of glucose solution against respective O.D. value of each.

Calculation

From the standard curve, the amount of glucose actually present in the sample is determined.

6.3 Total Sugar

Principle

The non-reducing sugars which are not titratable by Somogyi's reagent are at first hydrolyzed (broken down) to reducing sugars. Thus, after hydrolysis, the non-reducing sugars are converted to reducing sugars while the reducing sugars that are already present in the sample remain unchanged. Accordingly, all the sugars that are present after hydrolysis remain as reducing sugars. This is conveniently termed as total sugars.

Reagents

(A) to (H): Same as in reducing sugar.

(I) *Hydrochloric acid* (HCl): 5-10 per cent in distilled water (W/V).

(J) *Sodium hydroxide* (NaOH) Solution: 20-30 per cent in distilled water (W/V).

Procedure

1. *Extraction of sugars:* Extraction of sugars from the tissue sample is usually carried out by crushing with 80 per cent ethanol (Reagent-A). The crushed material is refluxed in a Soxhlet's apparatus (Figure 25) for about 30 minutes. After refluxing, alcohol is wiped out from the extract with the addition of distilled water. The extract is centrifuged. To the supernatant, saturated lead acetate solution (Reagent-C) and saturated di-sodium hydrogen phosphate solution (Reagent-D) are added. After centrifugation, the clear supernatant is collected.
2. *Hydrolysis:* For estimation of total sugars, the non-reducing sugars present in the extract are needed to be hydrolyzed at first. This may be conveniently done by the addition of any mineral acid. Thus, little amount of the acid (Reagent-I) is added to the extract and then it is warmed in a water-bath (50°–60° C) for few minutes for complete hydrolysis. When non-reducing sugars are hydrolyzed, the extract is cooled to room temperature and the *p*H of this is made neutral or slightly alkaline. To do so, sodium hydroxide solution (Reagent-J) is added drop by drop to the extract and its *p*H is also simultaneously tested with strips of *p*H or red litmus paper.
3. When the extract has come to neutral or slightly alkaline *p*H, it is made upto any known volume with distilled water.
4. Rest of the procedure remains same as in reducing sugar (Ref. 6.2: Procedure -Steps 2 to 8).

Calculation

Same as in case of reducing sugar (Ref. 6.2 : Calculation).

6.4 Non-reducing Sugar

The non-reducing sugars present in the sample may be determined from the values of the total and the reducing sugars as follows.

Percentage of non-reducing sugars =
[Percentage of (Total sugars) – (Reducing sugars)] × 0.95

6.5 Starch

Starch is a reserved form of polysaccharide in plants. Chemically, it is composed of 2 polysaccharides, which are amylose and amylopectin. Amylose is a straight-chain polymer of glucose units having α-1,4 glycosidic bonds. The other component, amylopection is a highly branched polymer of glucose with α-1,4 link and α-1,6 glycosidic bonds. Both these forms of polysaccharides yield α-D glucose units on hydrolysis.

Principle

The polysaccharide complex, starch is extracted with perchloric acid, after removing the soluble sugars present in the tissue sample. This is hydrolyzed into simple sugars and reacted with anthrone reagent to form green colouration. The intensity of the colour appeared is then measured calorimetrically.

Reagents

(A) *Ethanol* (C_2H_5OH): 80 per cent in distilled water (W/V).

(B) *Perchloric acid* ($HClO_4$): 50 per cent in distilled water (W/V).

(C) *Anthrone reagent:* 0.2 per cent [200 mg anthrone ($C_4H_{10}O$) is dissolved in little amount of ice-cold H_2SO_4 and the volume is made 4 upto 100 ml with the solvent].

(D) *Standard glucose solution:* 100 mg of glucose ($C_6H_{12}O_6$) is weighed, dissolved in little amount of distilled water and the volume is made upto 100 ml with distilled water. From this solution, 10 ml is taken and this is diluted to make 100 ml with distilled water. The concentration of this solution will be 100 μg/ml.

Procedure

1. Known amount of tissue sample by weight is taken. This is extracted with 80 per cent ethanol (Reagent-A) in a boiling water-bath for 1-2 hours. Then it is cooled to room temperature.
2. The extract is centrifuged and the residue is taken. The residue is then extracted with 50 per cent perchloric acid (Reagent-B) at 4°C temperature for 20 minutes.
3. After centrifugation, the supernatant is taken out and made upto a known volume with perchloric acid.

4. A number of aliquots (0.5-1.0 ml) are taken separately into test-tubes. Each should be made upto 1 ml with distilled water.
5. 4 ml of 0.2 per cent anthrone reagent (Reagent C) is then added to each test-tube and then heated in a boiling water-bath for about 10 minutes.
6. Test-tubes are cooled to room temperature. The green to dark-green colouration will appear on heating the samples. The Optical Density (O.D.) value of the coloured solutions is then measured through 630 nm wavelength in a colorimeter against blank.
7. *Preparation of standard curve:* From the standard glucose solution (100 μg/ml) (Reagent-D), different concentration gradients are made. To each of the solution, green colouration is developed by the same procedure. The O.D. value of it is then measured through 630 nm wavelength in a colorimeter against blank. A standard curve is prepared by plotting the known concentrations of glucose solution against respective O.D. value of each.

Calculation

From the standard curve, the amount of glucose actually present in the sample is determined. The starch content of the sample is calculated on multiplying the obtained value with 0.90.

6.6 Total Pectic Substances

Pectic substances are polyuronides and are composed mainly of 1,4 linked-α-D-galacturonic acid, or its methyl ester with neutral sugars, *i.e.*, galactose and arabinose as the side chains. The carboxyl groups of the galacturonic acid molecules are partly esterified with methanol and the free- groups are more or less neutralized. Pectic substances are located only in the middle-lamellae and the primary cell-wall in plant tissue.

Principle

After the extraction of pectin from plant tissue, it is saponified with alkali and precipitated as calcium pectate. After washing, the precipitate is dried and weighed.

Reagents

(A) *Hydrochloric acid* (HCl): 0.05 *N*.

(B) *Sodium hydroxide* (NaOH) *solution:* 1 *N* in distilled water (W/V).

(C) *Acetic acid* (CH_3COOH): 1 *N*.

(D) *Calcium chloride* ($CaCl_2$) *solution:* 1 *N* in distilled (W/V).

Procedure

1. Known amount of tissue sample by weight is taken. This is boiled with 0.05 *N* hydrochloric acid (Reagent-A) at a temperature of 80-90°C for 2 hours to extract the pectic substances.
2. After extraction, the mixture is cooled to room temperature. It is then filtered (Whatman No.4). The extraction is repeated for maximum recovery of the pectin.
3. The filtrate is made upto a known volume with distilled water.
4. An aliquot from the above extract (50 ml) is taken in a conical flask and it is neutralized with 1 *N* sodium hydroxide solution (Reagent-B). The excess amount of sodium hydroxide solution is added to it with thorough mixing till the mixture becomes slightly alkaline. The solution is allowed to stand overnight.
5. Then 10 ml of 1 *N* acetic acid (Reagent-C) is added to it. After 5 minutes, 5ml of 1 *N* calcium chloride solution (Reagent-D) is mixed with constant stirring. It is left for an hour to precipitate the calcium pectate.
6. The solution is then filtered through a dried and pre-weighed filter paper (Whatman No.4). The precipitate is washed repeatedly with boiling water to eliminate chloride ions present in the precipitate.
7. The filter paper containing calcium pectate is dried at 50°-60° C and weighed.

Calculation

The amount of total pectic substances (as calcium pectate) present in the sample is calculated by the following formula.

Percentage of pectic substances (as calcium pectate) in sample

$$= \frac{d \times c}{b \times a} \times 100$$

where,

a = Weight of sample.
b = Volume of aliquot taken for estimation.
c = Volume made with distilled water.
d = Weight of calcium pectate.

6.7 Type of Pectic Substances

Pectic substances are classified in several ways. According to the American Chemical Society (1926), these *complex polysaccharides* have been classified into 4 groups and these have been nomenclated as protopectin, pectinic acid, pectin and pectic acid.

Protopectin

The term, protopectin is applied to the water-insoluble parent pectic substances which occurs in plants and which, upon restricted hydrolysis, yields pectinic acid.

Pectinic acid

The term, pectinic acid is used for colloidal polygalacturonic acid containing more than a negligible proportion of methyl ester groups.

Pectin

The term, pectin designates those water-soluble pectinic acids of varying methyl ester content and degree of neutralization which are capable of forming gels with sugar acid under suitable condition.

Pectic acid

The term, pectic acid is applied to pectic substances mostly composed of colloidal polygalacturonic acids and essentially free from methyl ester groups.

However, the classification and fractionation of pectic substances in consideration to their *chemical solubility* is considered more acceptable in pectic chemistry at the present time. In this system, pectic substances are commonly classified into the following groups.

Water-soluble pectic substances

Under this group comes those peqtic substances, which are soluble in water and these include pectin and colloidal pectinic acids of high methyl ester content.

Oxalate-soluble pectic substances

These include pectic acid and colloidal pectinic acids of low methyl ester content.

Acid-soluble pectic substances and alkali-soluble pectic substances

These groups include protopectin, in other words, water-insoluble forms of pectic substances having high methyl ester content.

Principle

Pectic substances are fractionated using different types of extracting solvents. Each of the fracti... containing pectic substances is then reacted with sulphuric acid and *m*-hydroxybiphenyl solution to form a red coloured complex. The intensity of the colour appeared is then measured colorimetrically to determine the amount present in the sample.

Reagents

(A) *Ethanol* (C_2H_5OH): 90-95 per cent in distilled water (W/V).

(B) *Potassium oxalate* ($K_2C_2O_4$, H_2O) *solution:* 1 per cent in distilled water (W/V).

(C) *Hydrochloric acid* (HCl): 0.05 *M*.

(D) *Sodium hydroxide* (NaOH) *solution:* 0.05 *M* in distilled water (W/V).

(E) *Sodium tetraborate-sulphuric acid solution:* 0.0125 *M* [Sodium tetraborate ($Na_2B_4O_7$, 10 H_2O) is weighed, dissolved in pre-chilled H_2SO_4 and is made upto the required volume].

(F) *Sodium hydroxide* (NaOH) *solution:* 0.5 per cent in distilled water (W/V).

(G) *m-hydroxybiphenyl* ($C_{12}H_{10}O$) *solution:* 0.15 per cent in 0.5 per cent NaOH solution (W/V).

(H) *Standard galacturonic acid solution:* 120.5 mg of galacturonic acid monohydrate ($C_6H_{10}O_7$, H_2O) is weighed and taken in a volumetric flask. 5 ml of 0.05 *N* NaOH is added to it and the volume is made upto 100 ml with distilled water. From this solution, 10 ml taken and this is diluted to make 100 ml with distilled water. The concentration of this solution will be 100 μg/ml.

Procedure

1. *Preparation of alcohol-insoluble residue (AIR)*: Known amount of tissue sample by weight is taken. This is refluxed with ethanol (Reagent-A) in a Soxhelt's apparatus for 30 minutes. Then it is cooled to room temperature. The extract is filtered (Whatman No. 1) and washed repeatedly with ethanol till the material becomes colourless. The material is then dried at 40°C and weighed. This is considered as AIR.
2. From the AIR, different pectic fractions are extracted sequentially. Extraction should be done using thrice for each of the fraction.
3. Known amount of prepared AIR is taken and extracted with distilled water for 30 minutes at room temperature, dispersing well using a magnetic stirrer. The mixture with distilled water is centrifuged (3000 g) and the supernatant is collected. This is considered as the *water-soluble pectic substances (WSP)*.
4. The residue obtained from WSP is extracted with 1 per cent potassium oxalate solution (*p*H 4.5) (Reagent-B) for 30 minutes at room temperature, dispersing well using a magnetic stirrer. After centrifugation (3000 g), the supernatant is collected and this is considered as the *oxalate-soluble pectic substances (OXP)*.
5. The residue obtained from OXP is extracted with 0.05 *M* hydrochloric acid (Reagent-C) at 80°-90° C for 30 minutes, cooled to room temperature and centrifuged as above. The supernatant after centrifugation is collected and this is considered as the *acid-soluble pectic fraction (HP)*.
6. The residue obtained from HP is extracted with 0.05 *M* sodium-hydroxide solution (Reagent-D) at 4°C for 30 minutes and centrifuged as above. The supernatant is collected and the *p*H is adjusted to 4.5. This is considered as the *alkali-soluble pectic substances (OHP)*.
7. The WSP, OXP, HP and OHP are separately used for estimation of pectic substances as galacturonic acid.

8. A number of aliquots (0.5-1.0 ml) from each are taken into test-tube. Each should be made upto 1 ml with distilled water.
9. 6 ml of 0.0125 *M* sodium tetraborate-sulphuric acid solution (Reagent-E) is added to each test-tube and heated in a boiling water-bath for 5 minutes. Test-tubes are then cooled to room temperature.
10. 0.1 ml of 0.15 per cent *m*-hydroxybiphenyl solution (Reagent-G) is added to the test-tube and vigorously shaken to develop the red colouration. The Optical Density (O.D.) value of the coloured solution is measured through 520 nm wavelength in a colorimeter against blank.
11. *Preparation of standard curve:* From the standard galacturonic acid solution (100 µg/ml) (Reagent-H), different concentration gradients are made. To each of the solution, red colouration is developed by the same procedure. The O.D. value of it is then measured through 520 nm wavelength in a colorimeter against blank. A standard curve is prepared by plotting the known concentrations of galacturonic acid solution against respective O.D. value of each.

Calculation

From the standard curve, the amount of pectic substances as galacturonic acid present in the sample is determined.

6.8 Crude Fibre

Crude fibre is considered as the material left after making digestion of the tissue. It is mainly composed of cellulose, lignin and some minerals.

Principle

Cellulose and lignin in plant tissue are digested by reacting with acid and alkali. On filtration, the residue is obtained, which on drying is weighed. It is then ashed and reweighed. From the ash weight, the amount of crude fibre present in the sample is determined.

Reagents

(A) *Ether-ethanol mixture*: 1 part of di-ethyl ether (C_2H_5-O-C_2H_5) is mixed with 2 parts of ethanol (C_2H_5OH) and thoroughly homogenized.

(B) *Sulphuric acid* (H_2SO_4): 0.255 *N*.

(C) *Sodium hydroxide* (NaOH) *solution*: 0.313 *N*.

(D) *Sulphuric acid* (H_2SO_4): 0.05 *N*.

(E) *Ethanol* (C_2H^5OH): 70 per cent in distilled water (W/V).

Procedure

1. Known amount of tissue sample by weight is taken. This is extracted with ether-ethanol mixture (Reagent-A) and warmed (less than 50°C) to remove the lipid substances present. It is filtered and the residue is obtained.
2. Then it is boiled with 0.255 *N* sulphuric acid (Reagent-B) for 30 minutes. After boiling, the extract is filtered (Whatman No. 1) and the residue is washed repeatedly with boiling water to remove the acid.
3. It is boiled with 0.313 *N* sodium hydroxide (Reagent-C) for 30 minutes and then filtered. The residue is sequentially washed with dilute sulphuric acid (Reagent-D), distilled water and ethanol (Reagent-E).
4. The residue is then taken in a pre-weighed crucible (b) and dried at 100°-110° C for 2 hours. The crucible containing the residue is cooled and re-weighed (c).
5. The crucible is then ignited at about 600° C for few minutes in a muffle furnace. The crucible containing the ash is cooled and re-weighed (d).

Calculation

The amount of crude fibre present in the sample is calculated by the following formula.

$$\text{Percentage of crude fibre in the sample} = \frac{(c-b)-(d-b)}{a} \times 100$$

where,

a = Weight of sample.

b = Weight of crucible.

c = Initial weight of crucible containing tissue sample before ignition.

d = Final weight of crucible containing ash after ignition.

6.9 Total Soluble Solids (with a Refractometer)

Total soluble solids (TSS) of a given sample of fruit juice represent the various chemical substances present in it in soluble form. It indicates a measure of sugars present in the sample. The amount of TSS present in the juice of a fruit is also considered to be a reliable index in judging its maturity. In accordance with, the harvest-maturity of many fruits is assessed in considering the TSS of their juices.

Principle

The TSS of a given fruit juice sample is determined in a quicker way with the help of a refractometer, which is also known as hand- or pocket-refractometer (Figure 26). The instrument works on the principle of refractive index of the sample and gives the refractive index as °Brix.

Requirements

A hand-refractometer. A dropper or a glass-rod. Blotting paper. Absorbent cotton. Rectified spirit (if possible). Double (glass)-distilled water.*

Procedure

1. The lid, *i.e.*, covering plate of the refractometer which rests over the prism-plate and is attached with it at the basal end with a hinge is unfolded backward. By doing so, both the lid and the prism-plate are exposed.
2. The lid and the prism-plate are then carefully and scrupulously washed with jet of clean water to ensure that they have no stain on their surfaces.
3. Water adhered on the prism-plate and the lid as well as the surrounding parts of them, if any is completely wiped off with blotting paper or absorbent cotton.
4. The lid of the prism-plate is then washed with double (glass)- distilled water and the water adhered on them is blotted out. The cleaning is best done by rubbing the lid and the prism-plate gently and carefully with an absorbent cotton, soaked with rectified spirit.

---

* Instead of using ordinary distilled water which may contain traces of soluble substances, it is advisable to use double (glass)- distilled water, prepared with distilled water in all-glass retort.

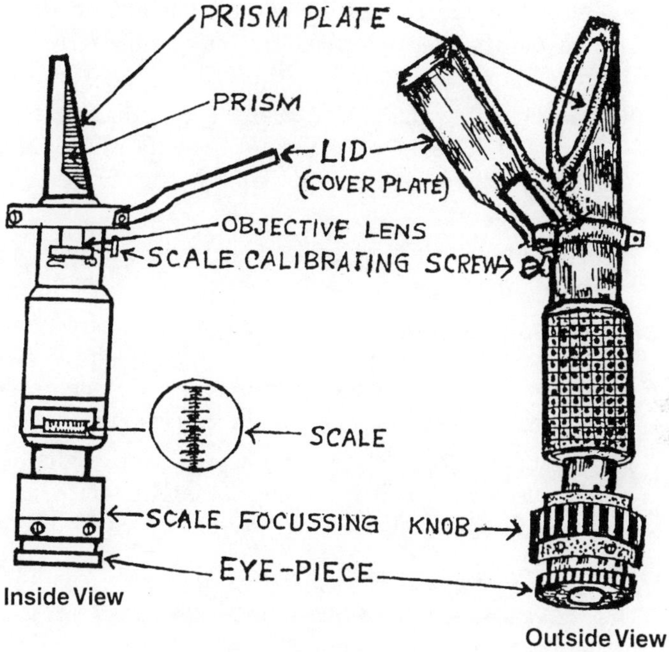

Fig. 26: Hand-refractometer.

5. Then, with the help of a previously cleaned dropper or a glass-rod, a drop of double (glass)- distilled water is carefully dropped on the surface of the prism-plate. The lid is folded forward and placed over the prism-plate to cover it. At this position, the lid and the prism-plate are firmly held together with fingers to avoid unfolding of the lid.
6. The refractometer is held to point it towards light. The eye-piece of it is brought close to any eye of the observer who should look into the eye-piece to have a view of the image of the scale. The scale-focussing knob should be conveniently rotated to adjust it at such a position where the scale is most clearly visible. The shaded part would be seen to intersect the unshaded part at *zero position* of the scale which indicates no reading with respect to distilled water. (The reading may not come to zero due to previous mishandling of the instrument. In such case, it should be

brought to zero by rotating the scale-calibrating screw. This may also happen if the distilled water contains any soluble matter).

7. The lid is then folded backward. The distilled water that remains adhered over the lid and the prism-plate is completely blotted out and these are dried in air for a few minutes.
8. The dropper or the glass-rod is also dried to make free from adhering water.
9. After that, a clear sample of fruit juice, TSS of which is to be determined is taken in the dropper, or a drop may be taken with the glass-rod. A drop of juice is, then carefully placed on the prism-plate, in the same way as in (5).
10. Rest of the procedure remains same as has been done with distilled water (6). The reading of the juice sample as °Brix is obtained and amount of TSS is expressed accordingly.

Correction factor

It should be noted that temperature is an important factor in the measurement of refractive index. The calibrating drop of distilled water should in fact, be at the same temperature as with the temperature of the juice sample. The determination should be done at 20°C. However, if determination is done at temperature which is not 20°C, the observed reading should be corrected to standard temperature of 20°C. For this, correction should be made in consideration to the following correction chart (Table 10).

6.10 Total Titratable Acidity*

Acids are important constituents in fruits as together with sugars, they determine quality and taste of the fruits. Maturity of many fruits for their harvest is also judged from their level of acids along with sugars, or the soluble solids. Fruits contain organic acids and among inorganic acids, only phosphoric acid is present. Acids that are accumulated in fruits are largely synthesized in leaves and are translocated to fruits.

In some fruits, one or more acids may be present in relatively high amount than the other acids and accordingly, these are referred

---

* Total titratable acidity refers to the amount of acidity present in *free + combined* state. *Free acidity* denotes the amount of acidity present in free state only while actual acidity refers to the amount of (H^+) ion concentration present.

Table 10: Temperature correction chart for refractometer reading

Temperature °C		Reading 0	5	10	15	20	25	30	35	40	45	50	55	60	65	70
10	Deduct from reading	0.50	0.54	0.58	0.61	0.64	0.66	0.68	0.70	0.72	0.73	0.74	0.75	0.76	0.78	0.79
11		0.46	0.46	0.53	0.55	0.58	0.60	0.62	0.64	0.65	0.66	0.67	0.68	0.69	0.70	0.71
12		0.42	0.45	0.48	0.50	0.52	0.54	0.56	0.57	0.58	0.59	0.60	0.61	0.60	0.63	0.63
13		0.37	0.40	0.42	0.44	0.46	0.48	0.49		0.51	0.52	0.53	0.54	0.54	0.55	0.55
14		0.33	0.35	0.37	0.39	0.40	0.41	0.42	0.43	0.44	0.45	0.45	0.46	0.46	0.47	0.48
15		0.27	0.29	0.31	0.33	0.34	0.34	0.35	0.36	0.37	0.37	0.38	0.39	0.39	0.40	0.40
16		0.22	0.24	0.25	0.26	0.27	0.28	0.28	0.29	0.30	0.30	0.30	0.31	0.31	0.32	0.32
17		0.17	0.18	0.19	0.20	0.21	0.22	0.21	0.22	0.22	0.23	0.23	0.23	0.23	0.24	0.24
18		0.12	0.13	0.13	0.14	0.14	0.14	0.14	0.15	0.15	0.15	0.15	0.16	0.16	0.16	0.16
19		0.06	0.06	0.06	0.07	0.07	0.07	0.07	0.08	0.08	0.08	0.08	0.08	0.08	0.08	0.08
20		0	0	0	0	0	0	0	0	0	0	0	0	0	0	0
21	Add to reading	0.06	0.07	0.07	0.07	0.07	0.08	0.08	0.08	0.08	0.08	0.08	0.08	0.08	0.08	0.08
22		0.13	0.13	0.14	0.14	0.15	0.15	0.15	0.15	0.15	0.16	0.l6	0.16	0.16	0.16	0.16
23		0.19	0.20	0.21	0.22	0.22	0.23	0.23	0.23	0.23	0.24	0.24	0.24	0.24	0.24	0.24
24		0.26	0.27	0.28	0.29	0.30	0.30	0.31	0.31	0.31	0.31	0.31	0.32	032	0.32	0.32
25		0.33	0.35	0.36	037	0.38	038	0.39	039	0.40	0.40	0.40	0.40	0.40	0.40	0.40
26		0.40	0.42	0.43	0.44	0.45	0.46	0.47	0.47	0.48	0.48	0.48	0.48	0.48	0.48	0.48
27		0.48	0.50	0.52	0.53	0.54	0.55	0.56	0.55	0.56	0.56	0.56	0.56	0.56	0.56	0.56
28		0.56	0.57	0.60	0.61	0.62	0.63	0.63	0.63	0.64	0.64	0.64	0.64	0.64	0.64	0.64
29		0.64	0.66	0.68	0.69	0.72	0.72	0.72	0.72	0.71	0.71	0.73	0.73	0.73	0.73	0.73
30		0.72	0.74	0.77	0.78	0.79	0.80	0.81	0:81	0.81	0.81	0.81	0.81	0.81	0.81	0.81

Source: International Sugar Analysis Control Committee.

to as the *predominant acids* respective to these fruits. For example, citric acid is the predominant acid in fruits like citruses and also in strawberry, currants etc., while malic acid is predominant in apple, cherry, plum etc. tartaric and malic acids in grapes (depending on season) and so on.

Principle

The total acidity of a fruit could be determined by titrating a known amount of aqueous extract of it against an alkali solution of known normality. It is expressed as *equivalence* of any organic acid, e.g., citric, malic etc.

Reagents

(A) *Sodium hydroxide* (NaOH) *solution*: Any known normality.

(B) *Phenolphthalein* ($C_{20}H_{14}O_4$) *indicator*: Approx. 0.5 per cent in 80 per cent ethanol (W/V).

Procedure

1. Known amount of tissue sample by weight is taken. This is extracted with distilled water by thorough crushing. Extract is filtered and the filtrate is made upto a known volume with distilled water.
2. A known volume of aliquot is taken in a conical flask. Few drops of phenolphthalein solution (Reagent-B) are added to this and shaken well.
3. A burette is filled with the sodium hydroxide solution (Reagent-A) after washing and rinsing.
4. Titration is done and the end-point is determined by the appearance of pink colour and its persistence for at least few seconds.

Calculation

In terms of citric acid

1 ml of 0.1 N NaOH solution can neutralize 0.064 g of citric acid. Therefore, percentage of total titratable acidity in the sample as equivalence of citric acid =

$$\frac{d \times 0.064 \text{ (if NaOH solution is 0.1 } N) \times C}{a \times b} \times 100$$

where,

a = Weight of sample.
b = Volume of aliquot taken for examination.
c = Volume made with distilled water.
d = Average burette reading for sample.

6.11 Vitamic C

Vitamin C or ascorbic acid is an enediol isomer of 2-keto-L-gluconolactone with a configuration similar to that of L-glucose. Oxidation of ascorbic acid gives rise to dehydro-ascorbic acid and both forms are physiologically active.

Principle

Titrimetric estimation of vitamin C is conventionally done using 2,6-dichlorophenol indophenol dye solution. This dye is blue in alkaline solution and red in acidic solution. Ascorbic acid reduces the dye to a colourless form. Reaction is quantitative and specific for ascorbic acid at *p*H 1.0 -3.5.

Reagents

(A) *Metaphosphoric acid* (HPO_3): 3 per cent in distilled water (W/V).

(B) *2,6-dichlorophenol indophenol* ($C_{12}H_6C_2NNaO_2$, $2H_2O$) *solution*: 0.025 per cent [50 mg 2,6-dichlorophenol indophenol is weighed and taken in a beaker. 42 mg sodium bicarbonate (Na_2HCO_3) is added to it. It is then dissolved in little amount of distilled water and the volume is made upto 200 ml with distilled water].

(C) *Standard ascorbic acid* ($C_6H_8O_6$) *solution:* 0.01 per cent in 3 per cent metaphosphoric acid (W/V).

Procedure

1. Known amount of tissue sample by weight is taken. This is extracted with 3 per cent metaphosphoric acid (Reagent-A) by thorough crushing. Extract is filtered and filtrate is made upto a known volume with 3 per cent metaphosphoric acid.
2. A known volume of aliquot is taken in a conical flask. A burette is filled with indophenol dye solution (Reagent-B) and titration is done. End-point is determined by the

appearance of pink colour which should persist for at least 15 seconds.

3. *Standardization of the dye solution*: The dye solution is needed to be standardized simultaneously. For this, 5 ml of standard ascorbic acid solution (Reagent-C) is taken in a conical flask and to this, 5 ml of 3 per cent metaphosphoric acid solution is added. Both are thoroughly mixed by shaking and is titrated against indophenol dye solution to get the pink colour end-point and its persistence for few seconds.

Calculation

The *dye factor* (d) =

$$\frac{0.5}{\text{Average burette reading for standardization of dye solution}}$$

Vitamin C content in the sample is determined by the following formula.

$$\text{Vitamin C (mg per 100 g)} = \frac{e \times d \times b}{c \times a} \times 100$$

where,

a = Weight of sample.

b = Volume made with metaphosphoric acid.

c = Volume of aliquot taken for estimation.

d = Dye factor.

e = Average burette reading for sample.

6.12 Total Free– Amino Acids

Amino acids are organic acids that contain one or more amino and carboxyl groups and thus, they possess both acidic and basic properties. Most of the amino acids are crystalline and nearly all of them are α-amino acids, in that the amino group is present on the α-carbon atom.

The general formula of an amino acid is R-CH-NH_2-COOH, where R varies according to the type of the amino acids Amino acids that are present as units of plant proteins are 20 in number, which are glycine, alanine, valine, leucine, isoleucine, serine, threonine, cysteine, cystine, methionine, aspartic acid, glutamic acid, lysine, arginine, phenylalanine, tyrosine, tryptophan, histidine, proline and hydroxyproline.

(Amide is an amino ester of an amino acid and results when OH of the COOH group of an amino acid is replaced by an NH_2 group. Best known amides in plants are asparagine and glutamine, which are amino estrs of aspartic and glutamic acids respectively.)

Principle

The solution of free-amino acids reacts with ninhydrin to give rise to blue colouration. (The blue colouration is obtained by all amino acids having a free- α-NH_2 group and thus, not by proline). The intensity of the blue colour obtained by reacting with ninhydrin is determined to get the amount of free– amino acids present in the sample.

Reagents

(A) *Ethanol* (C_2H_5OH): 80 per cent in distilled water (W/V).

(B) *Ninhydrin solution:* 0.5 g ninhydrin ($C_2H_6O_4$) and 0.05 g hydrindantin ($C_{18}H_{10}O_6$) are dissolved in 12.5 ml of 60 per cent methyl cellosolve ($C_3H_8O_2$). To this, 4 ml of 4 *M* acetate buffer (*p*H 5.4) is added and volume is made upto 100 ml with methyl cellosolve.

(C) *Standard amino acid solution:* 100 mg of glycine ($C_2H_5NO_2$) is dissolved in little amount of distilled water and the volume is made upto 100 ml with distilled water. From this solution, 10 ml is taken and this is diluted to make 100 ml with distilled water. The concentration of this solution will be 100 µg/ml.

Procedure

1. Known amount of tissue sample by weight is taken. This is extracted with 80 per cent ethanol (Reagent-A) at 70°-80° C by refluxing in a Soxhlet's apparatus for an hour. The extract is cooled to room temperature.
2. After cooling, the extract is filtered (Whatman No.1) or centrifuged (4000 g). The supernatant is collected and made upto a known volume with distilled water.
3. An aliquot (1 ml) is taken in a test-tube. To this, 2 ml of 0.5 per cent ninhydrin solution (Reagent-B) is added, shaken

well and heated in a water-bath (70°-80° C) for 10 minutes, placing suitable lid.

4. The solution developing blue colour is allowed to cool to room temperature. This is made upto a known volume with distilled water.
5. The Optical Density (O.D.) value of the coloured solution is then measured through 590 nm wavelength in a colorimeter against blank.
6. *Preparation of standard curve*: From the standard glycine solution (100 μg/ml) (Reagent-C), different concentration gradients are made. To each of the solution, blue colouration is developed by the same procedure. The O.D. value of it is then measured through 590 nm wavelength in a colorimeter against blank. A standard curve is prepared by plotting the known concentrations of glycine solution against respective O.D. value of each.

Calculation

From the standard curve, the amount of total free- amino acids, present in the sample is determined and expressed as equivalence of glycine.

6.13 Separation and Detection of Free– Amino Acids (by Thin Layer Chromatography)

Principle

The different free– amino acids that are present in fruits may be separated out in a simpler way by the technique of two-dimensional chromatography. When separated on the chromatogram, the locations of the individual amino acids may be detected as coloured spots or specks which are produced on reaction with ninhydrin. The amino acids represented by the coloured spots on the chromatogram may be identified by comparing with coloured spots developed on another chromatogram prepared with solutions of known amino acids.

Reagents

(A) *Sodium acetate* ($NaC_2H_3O_2$, $3H_2O$) *solution:* 0.02 *M* in distilled water (W/V).

(B) *Silica gel G*: Slurry to be prepared with 0.02 *M* sodium acetate solution.

(C) *Ethanol* (C_2H_5OH): 80 per cent in distilled water (V/V).

(D) *Solvent 1*: 80 per cent phenol (C_6H_6O).

(E) *Solvent 2*: *n*-butanol ($C_4H_{10}O$), glacial acetic acid ($C_2H_4O_2$) and distilled water, respectively in the proportion of 4, 1 and 1 part by volume are mixed together and thoroughly homogenized.

(F) *Isopropanol* (C_3H_8O): 10 per cent in distilled water (V/V).

(G) *Standard solutions of amino acids*: 100 µg/ml in 10 per cent isopropanol (W/V).

(H) *Ninhydrin solution*: (Same as in 6. 12 → Reagent B).

Procedure

1. *Preparation of thin layer plates*: The gla.s-plates of convenient size(s) used for the purpose are at first thoroughly cleaned with rectified spirit and dried. Then aqueous slurry of silica gel (Reagent-B) is spread at one side over each plate with the help of a suitable spreader. The thickness of slurry on each plate is adjusted to 250 µm. Then the plates are dried and activated on exposing to 100°C for an hour in an oven. After that, these are brought outside and cooled to room temperature. Two lines are then drawn on each plate with a lead pencil in right angles near and parallel to two edges of it so that these intersect one point.
2. Known amount of tissue sample by weight is taken. This is extracted with 80 per cent ethanol (Reagent-D) by refluxing in a Soxhlet's apparatus for an hour. The extract is cooled to room temperature.
3. After cooling, the extract is filtered (Whatman No.1) or centrifuged (4000 g) and the supernatant is collected.
4. The supernatant is concentrated in an evarporator.
5. At the intersecting point of the two lines on a plate, 20- 50 µl of the supernatant is carefully spotted drop by drop with a micro-pipette and dried simultaneously with a hand-drier.
6. The plate, thus applied with the extract is suitably placed in a chromatography chamber, which is pre-saturated with Solvent-1 (Reagent-D). The Solvent-1 is then allowed to run on the plate upto a desired distance, avoiding running

very close to the edge of the plate. The plate is then carefully brought outside the chamber and dried in an oven. After this, it is brought outside and cooled to room temperature.

7. On a separate plate applied with silica gel, an aliquot of about 20 µl respectively for each of the amino acid solution (Reagent-G) is spotted side by side maintaining suitable distances. This is done in a similar manner as in step 5. (If all amino acids are not possible to be spotted on one plate, more than one plate may be used).
8. The plate(s) is then suitably placed in another chromatography chamber, pre-saturated with Solvent-2 (Reagent-E). The Solvent-2 is then allowed to run on the plate in a similar manner as in step 6. The plate(s) is dried similarly as in step 6.
9. The chromatograms, both, for the extract and solutions of known amino acids are then sprayed with ninhydrin solution (Reagent-H) so as to make complete coverage over them. After spraying, these are dried in a suitable oven for some time to develop colour and then brought outside and cooled to room temperature. Migration of each amino acid on the plate, thereby indicating its location on the chromatogram would be visible by the appearance of purple colour, while yellow coloured spot if any, would represent the locations of the amino acids, proline or hydroxy-proline.
10. *Identification of amino acids*: The amino acids on the chromatograms which have appeared as coloured spots may be identified from the *Rf* values of them. The *Rf* value for each spot, representing each amino acid developed on the chromatograms for the sample or for the known solutions of amino acids may be worked out as follows.

$$Rf = \frac{\text{Distance of the centre of the spot from origin}}{\text{Distance moved by the solvent from origin}}$$

The *Rf* values for the coloured spots appeared on the chromatogram developed in respect of application of the extract are to be compared with the *Rf* values of the spots appeared on the chromatogram(s) developed, in respect of application of known solution of amino acids to detect the amino acids present in the extract.

6.14 Protein

The amino acids are joined together by the peptide bonds (-CO-NH-), formed by the condensation of α-COOH of one amino acid with the α-NH_2 group of another. The condensed polymers containing upto 50 amino acid residues with molecular weights of less than 5000 are referred to as *peptides* while the term, protein is restricted to very large polymers of amino acids, having molecular weights of several thousands or more.

Proteins are the principal constituents of cell protoplasm. The main function of protein is to act as essential components of structural material. According to solubility and physico-chemical properties, proteins are usually classified into 3 broad groups, *i.e.*, simple proteins, which contain only amino acids, e.g., protamines, histones, globulins, albumins etc., conjugated proteins, which apart from amino acids contain phosphorus, carbohydrates, nucleic acids, *e.g.*, chromoproteins, nucleoproteins, glycoproteins, lipoproteins etc. and derived proteins which are the degraded products of proteins, *e.g.*, proleans, meta-proleins, proteoses etc.

Principle

Protein reacts with Folin and Ciocalteu's phenol reagent to give rise to blue coloured complex. The colouration is formed due to reaction of alkaline copper of the reagent with protein and the reduction of phospho-molybdate by the amino acids, tyrosine and tryptophan, present in the protein molecules. The intensity of the developed colouration of the protein solution depends on the amount of these aromatic amino acids present and hence, it varies for different proteins.

Reagents

(A) *Sodium hydroxide* (NaOH) *solution*: 0.1 *N* in distilled water.

(B) *Trichloroacetic acid* ($C_2HCl_3O_2$): 10 per cent in distilled water (W/V).

(C) *Copper sulphate* ($CuSO_4$, $5H_2O$) *solution*: 0.5 per cent in distilled water (W/V).

(D) *Sodium potassium tartrate* ($KNaC_4H_4O_6$, $4H_2O$): 1 per cent in distilled water (W/V).

(E) *Sodium carbonate* (Na_2CO_3) *solution*: 1 per cent in 1 *N* NaOH (W/V).

(F) *Protein reagent*: The reagent is freshly prepared in the proportion of 5 ml, 5 ml and 250 ml of Reagents-C, D and E respectively.

(G) *Folin and Ciocalteu's phenol reagent*: 1 *N*.

(H) *Standard protein solution*: 100 mg of bovine serum albumin (BSA) is dissolved in little amount of distilled water and the volume is made upto 100 ml with distilled water. The concentration of this solution will be 1 mg/ml.

Procedure

1. Known amount of tissue sample by weight is taken. This is extracted with 0.1 *N* sodium hydroxide solution (Reagent-A) for 30 minutes. The extract is centrifuged (1500 g) and the supernatant is collected.
2. 10 per cent trichloroacetic acid solution (Reagent-B) is added to this at the ratio of 1 : 2. Protein will be seen to have precipitated and the precipitated protein is recovered by centrifugation (1500 g) for 10 minutes.
3. The protein pellet is dissolved in 0.1 *N* sodium hydroxide solution (Reagent-A) and made upto a known volume with the same solvent.
4. An aliquot (1 ml) is taken in a test-tube and 4 ml of protein reagent (Reagent-F) is added to this. It is then incubated at room temperature for 10 minutes.
5. 0.5 ml of Folin & Ciocalteu's phenol reagent (Reagent-G) is added to the test-tube to develop blue colouration in the solution Allowing for 10 minutes, the Optical Density (O.D.) value of the coloured solution is measured through 750 nm wavelength against blank.
6. *Preparation of standard curve*: From the standard protein solution (1 mg/ml) (Reagent-H), different concentration gradients are made. To each of the solution, blue colouration is developed by the same procedure. The O.D. value of it is then measured through 750 nm wavelength in a colorimeter against blank. A standard curve is prepared by plotting the known concentrations of protein solution against respective O.D. value of each.

Calculation

From the standard curve, the amount of protein present in the sample is determined.

6.15 Lipid (Ether Extractive)

Lipids are fatty materials which are insoluble in water but soluble in non-polar **solvents**, *i.e.*, ether, chloroform, benzene etc. Alkaline hydrolysis (saponification) of lipids gives rise to alcohol and sodium or potassium salts of the constituent fatty acids. Many classifications of lipids have been proposed on the basis of their structural, physico-chemical, biological or physiological properties. Chemically, lipids can be divided into 2 broad groups, viz., simple lipids and compound lipids. However, a third group is also there, referred to as derived lipids, *e.g.*, steroids.

Principle

Lipids are extracted in non-polar solovents. On extraction and subsequent evaporation of the solvent, the lipid extractive remains as residue and this is determined by weight.

Reagents

(A) *Ether-ethanol mixture*: 1 part of di-ethyl ether (C_2H_5–O–C_2H_5) is mixed with 2 parts of ethanol (C_2H_5OH) and thoroughly homogenized.

Procedure

1. Known amount of tissue sample by weight is taken. The sample is made into pieces and crushed with ethanol.
2. Crushed material is then refluxed with ether-ethanol mixture (Reagent-A) in a Soxhlet's apparatus not exceeding a temperature of 50°C for about 3-4 hours. Then it is cooled to room temperature and centrifuged.
3. The supernatant containing the lipid matter is then taken in a pre-weighed, suitable flat-bottom glass container with rim (b). The container with material (c) is weighed and is suitably warmed at a temperature below 50°C to evaporate the solvent. Heating is continued till evaporation is complete.
4. After cooling to room temperature, the container with the material (d) is weighed.

Calculation

The amount of lipid matter present in the sample is calculated by the following formula.

$$\text{Percentage of ether-extractive in sample} = \frac{(c-b) - (d-b)}{a} \times 100$$

where,

a = Weight of sample.

b = Weight of glass container.

c = Weight of glass container with material before evaporation of solvent.

d = Weight of glass container with material after evaporation of solvent.

6.16 Phenolic Compounds

Phenols are substances of aromatic compounds with hydroxyl groups. The simplest form of phenol, *i.e.*, catechol is not largely found and the most abundant phenolic compound is hydroquinone.

Principle

In alkaline medium, phenols reacts with Folin and Ciocalteu's phenol reagent to give rise to blue coloured complex. The intensity of the colour appeared is then measured colorimetrically.

Reagents

(A) *Ethanol* (C_2H_5OH): 80 per cent in distilled water (W/V).

(B) *Folin and Ciocalteu's phenol reagent*: 1 N.

(C) *Sodium carbonate* (Na_2CO_3) *solution*: 35 per cent in distilled water (W/V).

(D) *Standard phenol solution*: 100 mg of catechol ($C_6H_6O_2$) is dissolved in little amount of distilled water and the volume is made upto 100 ml with distilled water. From this solution, 10 ml is taken and this is diluted to make 100 ml with distilled water. The concentration of this solution will be 100 µg/ml.

Procedure

1. Known amount of tissue sample by weight is taken. This is extracted with 80 per cent ethanol (Reagent-A) in a boiling water-bath. It is then cooled to room temperature.
2. After cooling, the extract is centrifuged. The supernatant is taken and evaporated to nearly dryness in a water-bath.
3. The residue is dissolved in distilled water and made upto a known volume with distilled water.

4. A number of aliquots (0.5-1.0 ml) are taken separately into test-tubes. Each should be made upto 1 ml with distilled water.
5. 0.5 ml of Folin and Ciocalteu's phenol reagent (Reagent-B) is then added to each test-tube. After 5 minutes, 1 ml of 35 per cent sodium carbonate (Reagent-C) solution is added to this. The Optical Density (O.D) value of the coloured solution is then measured through 650 nm wavelength in a colorimeter against blank.
6. *Preparation of standard curve*: From the standard phenol solution (100 µg/ml) (Reagent-D), different concentration gradients are made. To each of the solution, the blue colour is developed by the same procedure. The O.D. value of it is then measured through 650 nm wavelength in a colorimeter against blank. A standard curve is prepared by plotting the known concentrations of phenol solution against respective O.D. value of each.

Calculation

From the standard curve, the amount of phenol actually present in the sample is determined.

6.17 Tannin

Tannins are compounds of high molecular weight (500- 3000), containing a large number of phenolic hydroxyl or other groups (1-2 per 100 mol. wt.), which enable them to form cross-links with proteins and some other compounds. For this property to from cross-links with proteins or other polymers, they inhibit enzyme systems. Tannins exist as hodrolyzable or condensed forms and due to presence in many fruits, especially at immature condition make them astringent in taste.

Principle

In alkaline medium, tannins react with Folin-Denis reagent to give rise to blue coloured complex. The intensity of the colour appeared is then measured colorimetrically.

Reagents

(A) *Folin-Denis reagent*: 50 g sodium tungstate (Na_2WO_4, $2H_2O$), 5 g phosphomolybdic acid and 25 ml of 85 per cent phosphoric acid (H_3PO_4) are dissolved in 300 ml of distilled water. The mixture is refluxed for 2 hours, cooled to 25°C and diluted to make 500 ml with distilled water.

(B) *Sodium carbonate* (Na_2CO_3) *solution*: 35 per cent in distilled water (W/V).

(C) *Standard tannic acid solution*: 100 mg of tannic acid is dissolved in little amount of distilled water and the volume is made upto 100 ml with distilled water. From this, 10 ml is taken and this is diluted to make 100 ml with distilled water. The concentration of this solution will be 100 µg/ml.

Procedure

1. Known amount of tissue sample by weight is taken. This is extracted with distilled water by thorough crushing.
2. Crushed material with distilled water is then boiled at a temperature of 100°C for 30 minutes. Then it is cooled to room temperature and centrifuged.
3. The supernatant is collected and made upto a known volume with distilled water.
4. A number of aliquots (0.1-1.0 ml) are taken separately into test-tubes. Each should be made upto 1 ml with distilled water. To this, 7.5 ml of distilled water and 0.5 ml of Folin-Denis reagent (Reagent-A) are added. After 1 minute, 1.0 ml of 35 per cent sodium carbonate solution (Reagent-B) is added. The reaction mixture is vigorously shaken and kept for 30 minutes to develop colouration at room temperature. The Optical Density (O.D.) value of it is then measured through 760 nm wavelength in a colorimeter against blank.
5. *Preparation of standard curve*: From the standard tannin solution (100 µg/ml) (Reagent-C), different concentration gradients are made. To each of the solution, the colour is developed by the same procedure. The O.D. value of it is then measured through 760 nm wavelength in a colorimeter against blank. A standard curve is prepared by plotting the known concentrations of tannin solution against respective O.D. value of each.

Calculation

From the standard curve, the amount of tannin actually present in the sample is determined.

6.18 Nitrogen (By Micro-Kjeldahl Method)

Nitrogen is a constituent of proteins, peptides, amino acids, amides, purines, pyrimidines, enzymes, co-enzymes, chlorophyll, porphyrins etc. in plants. Most of the higher plants take up nitrogen from the soil in the form of nitrate, which is sequentially reduced in the plant body for use in various anabolic reactions.

Principle

The various nitrogenous compounds present in the plant tissue are at first reacted with H_2SO_4 to form ammonium sulphate, $[(NH_4)_2SO_4]$. The ammonium sulphate, thus formed is then made to decompose by the action of an alkali solution and thereby, the ammonia (NH_3) gas is liberated. This is absorbed in sulphuric acid. The amount of ammonia liberated and absorbed in the acid is determined by titrating against a standard solution of an alkali.

Reagents

(A) *Digestion mixture*: 1.0 g copper sulphate ($CuSO_4$, $5H_2O$) and 2.0 g selenious acid ($H_2\ SeO_3$) are ground. These are added with 25 ml of orthophosphoric acid (H_3PO_4). The mixture is carefully shaken to dissolve the ingredients. To the solution, 75 ml of concentrated sulphuric acid (H_2SO_4) is added. The solution is homogenized and stored.

(B) *Sodium hydroxide* (NaOH) *solution*: 50 per cent in distilled water (W/V).

(C) *Sulphuric acid* (H_2SO_4): 0.5 *N*.

(D) *Sodium hydroxide* (NaOH) *solution*: 0.1 *N* or of any normality in distilled water.

(E) *Wislow's indicator*: 0.06 g of methyl red ($C_{15}H_{15}N_3O_2$) and 0.04 g of methylene blue ($C_{16}H_{18}ClN_3S$, $2H_2O$) are weighed and dissolved in 50 ml of 90-95 per cent ethanol.

Procedure

1. Known amount of dried tissue sample by weight is taken in a digestion flask (Figure 27a). This is digested (Figure 27 b) with digestion mixture (Reagent-A) till the material turns colourless. Then it is cooled to room temperature.
2. The digested material is transferred to the flask of the distillation set (Figure 27c). To this, 15 ml of 50 per cent NaOH solution (Reagent-B) is carefully added.

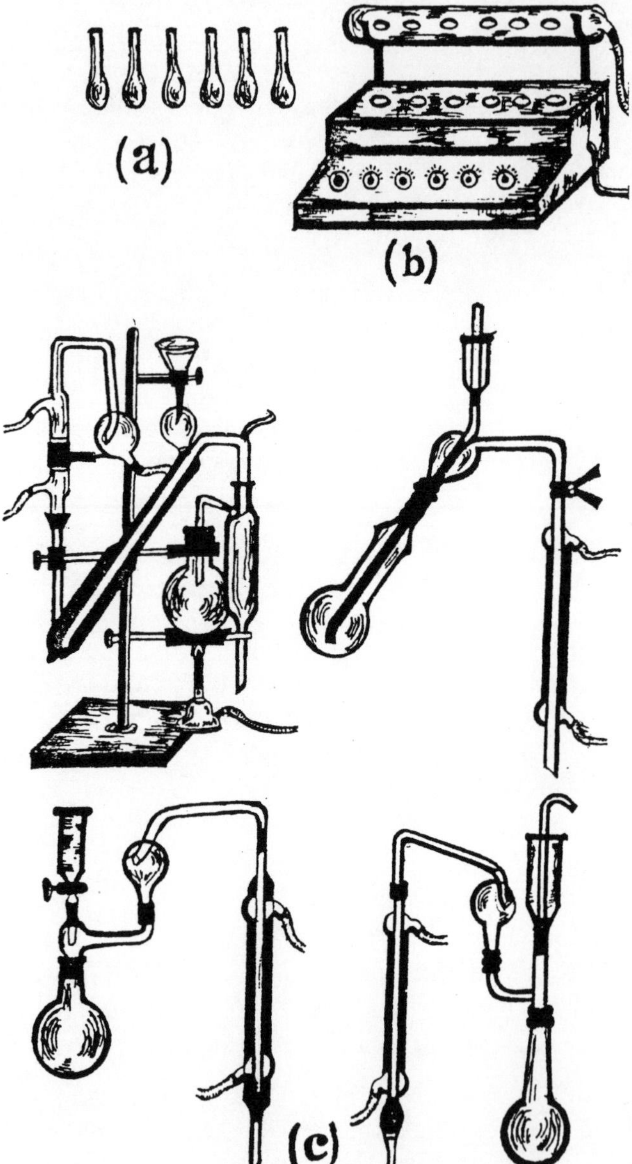

Fig. 27: Kjeldahl set.
(a) Digestion flasks. (b) Heater.
(c) Distillation apparatus of various types.

3. In a conical flask, 25 ml of 0.5 *N* sulphuric acid (Reagent-C) is taken. This should be placed below the outlet tube of distillation set, taking care that the outlet tube remains dipped in the 0.5 *N* H_2SO_4 taken in the conical flask.
4. Distillation is started by introducing steam. With progress of distillation, ammonia gas that is liberated will be collected and absorbed in the 0.5 *N* H_2SO_4 (Reagent-C) taken in the conical flask.
5. When distillation is completed, it is brought outside the distillation apparatus and cooled to room temperature. To this, 2-3 drops of Wislow's indicator (Reagent-E) are added to the conical flask. Colour will be purplish.
6. This is titrated against 0.1 *N* NaOH solution (Reagent-D) taken in a rinsed burette. End-point is determined by appearance of bluish green colour.
7. A blank sample should also be maintained side by side (containing no tissue) and titration of this should be done similarly.

Calculation

The amount of nitrogen present in the sample is calculated by the following formula.

1 ml of 0.1 N NaOH solution is equivalent to 0.014 g of nitrogen.

Therefore, percentage of nitrogen in the sample =

$$\frac{(b-c) \times d \times 0.014}{a} \times 100$$

where,

a = Weight of sample.
b = Burette reading for blank.
c = Burette reading for sample.
d = Normality of NaOH solution.

N.B. **Determination of protein content**

With consideration to the fact that plant proteins contain 16 per cent nitrogen, percentage of protein may be determined by multiplying nitrogen percentage with 6.25.

6.19 Phosphorous

Phosphorus is a constituent of nucleic acids, phospholipids, NADP and high energy compounds like ADP and ATP. The element is taken up by plants from the soil where it is present both in inorganic and organic forms.

Principle

Phosphorus is estimated by reacting with molybdic acid to form phosphomolybdate complex. It is reduced to form a blue coloured solution of molybdenum complex on reaction with amino-naphtholsulphonic acid. The intensity of the colour appeared is then measured colorimetrically.

Reagents

(A) *Digestion mixture*: 20 ml of nitric acid (HNO_3) and 2 ml of sulphuric acid (H_2SO_4) are taken and 8 ml of 70 per cent perchloric acid ($HClO_4$) is added slowly to it. The mixture is carefully shaken to mix well.

(B) *Sodium bisulphite* ($NaHSO_3$) *solution*: 15 per cent in distilled water (W/V).

(C) *Sodium sulphite* (Na_2SO_3) *solution*: 20 per cent in distilled water (W/V).

(D) *Fiske and Subbarow reagent*: 0.5 g 1-amino, 2-naphthol, 4-sulphonic acid [$NH_2C_{10}H_5(OH)_3SOH$] is taken and 5 ml of 20 per cent Na_2SO_3 solution (Reagent-C) is added to this. About 150 ml of 15 per cent $NaHSO_3$ solution (Reagent-B) is also added and warmed. Then it is cooled to room temperature and the volume is made upto 200 ml with $NaHSO_3$ solution.

(E) *Ammonium molybdate* [$(NH_4)_6 Mo_7O_{24}$, $4H_2O$] *solution*: 2.5 per cent in distilled water (W/V).

(F) *Sulphuric acid* (H_2SO_4): 10 *N*.

(G) *Standard phosphorus solution*: 0.438 g potassium dihydrogen phosphate (KH_2PO_4) is weighed and is dissolved in little amount of distilled water. 10 ml of 10 *N* H_2SO_4 is added to this and the volume is made upto 1 litre with distilled water. From this solution, 10 ml is taken and this is diluted to make 50 ml with distilled water. The concentration of this solution will be 100 µg/ml.

Procedure

1. Known amount of tissue sample by weight is taken in a crucible with lid. It is then ashed in a muffle-furnace. (N.B. Temperature should not exceed 600°C when ashed for estimation of phosphorus).
2. The ash is taken in a digestion flask. This is heated with digestion mixture (Reagent-A) till the material turns colourless. Then it is cooled to room temperature.
3. The digested material is then made upto a known volume with distilled water.
4. A number of aliquots (0.5-1.0 ml) are taken separately into test-tubes. Each should be made upto 1 ml with distilled water.
5. 2 ml of 2.5 per cent ammonium molybdate solution (Reagent-E) and 1 ml of Fiske and Subbarow reagent (Reagent-D) are added to each test-tube and allowed to stand for 10 minutes for development of colour. (If necessary, the test-tube may be slightly warmed to develop colour).
6. The blue colouration will appear due to reaction with phosphorus which is present in the sample. The Optical Density (O.D.) value of the coloured solution is then measured through 640 nm wavelength in a colorimeter against blank.
7. *Preparation of standard curve*: From the standard phosphorus solution (100 μg/ml) (Reagent-G), different concentration gradients are made. To each of the solution, the blue colour is developed by the same procedure. The O.D. value of it is then measured through 640 nm wavelength in a colorimeter against blank. A standard curve is prepared by plotting the known concentrations of phosphorus solution against respective O.D. value of each.

Calculation

From the standard curve, the amount of phosphorus actually present in the sample is determined.

6.20 Potassium

Although potassium is not a constituent of any compound in the plant body, it has an indispensable role for many metabolic reactions and also in opening and closure of stoma. It also seems to be involved in carbohydrate metabolism. The element is present in the soil in fixed, exchangeable or soluble forms and is taken up by many crop plants in relatively large amount (*luxury consumption*).

Principle

In estimating potassium, it is at first atomized by oxy-hydrogen or oxy-acetylene flames. The atoms of potassium which are excited by the flame, emit radiation at specific wavelength. The amount of radiation emitted is measured in a flame-photometer.

Reagents

(A) *Digestion mixture*: 20 ml of nitric acid (HNO_3) and 2 ml of sulphuric acid (H_2SO_4) are taken and 8 ml of 70 per cent perchloric acid ($HClO_4$) is slowly added to it. The mixture is carefully shaken to mix well.

(B) *Standard potassium solution*: 1.9 g potassium chloride (KCl) is weighed and is dissolved in little amount of distilled water. 5 ml of concentrated hydrochloric acid (HCl) is added to it and the volume is made upto 1 litre with distilled water. From this solution, 15 ml is taken and this is diluted to make 100 ml with distilled water. The concentration of this solution will be 150 μg/ml.

Procedure

1. The preparation of ash and digestion of it is done by the same procedure and is done in estimation of phosphorus (Ref. 6.19 : Procedure → Steps 1-2).
2. The digested material is made upto a known volume with distilled water.
3. An aliquot of digested material is atomized in a pre-calibrated flame-photometer.
4. The percentage of luminosity is then measured through 768 nm wavelength in a flame-photometer against blank.
5. *Preparation of standard curve*: From the standard potassium solution (Reagent-B), different concentration gradients are made. The percentage of luminosity is then measured through 768 nm wavelength in a flame-photometer against blank. A standard curve is prepared by plotting the known concentrations of potassium solution, against respective luminosity value of each.

Calculation

From the standard curve, the amount of potassium actually present ill the sample is determined.

6.21 Calcium

Calcium has its essentiality to exist as calcium-pectate, present in the cell-wall of a plant. It also acts as an activator of many enzymes. The plant takes up calcium from soil where it exists in non-exchangable form.

Principle

Titrimetric estimation of calcium is done by precipitating it as calcium oxalate. The precipitate is then dissolved in sulphuric acid and the amount of calcium dissolved in the acid is determined by titrating against a standard potassium permanganate solution.

Reagents

(A) *Digestion mixture*: 20 ml of nitirc acid (HNO_3) and 2 ml of sulphuric acid (H_2SO_4) are taken and 8 ml of 70 per cent perchloric acid ($HClO_4$) is added slowly to it. The mixture is carefully shaken to mix well.

(B) *Ammonium oxalate* [$(NH_4)_2 C_2O_4, H_2O$] *solution*: Saturated in distilled, water (W/V).

(C) *Methyl red* ($C_{15}H_{14}N_3O_2$) *indicator*: 0.5 per cent in 80 per cent ethanol (W/V).

(D) *Potassium permanganate* ($KMnO_4$) *solution*: 0.01 *N* in distilled water (W/V).

Procedure

1. The preparation of ash and digestion of it is done by the same procedure as is done in estimation of phosphorus (Ref. 6.19 : Procedure → Steps 1-2).
2. The digested material is made upto a known volume with distilled water.
3. The digested material is taken in a conical flask. To this, 10 ml of saturated ammonium oxalate solution (Reagent-B) and a few drops of methyl red indicator (Reagent-C) are added and shaken well.
4. With addition of ammonia (in liquor), the mixture is made slightly alkaline and it is turned to acidic state (*p*H 5.0) by adding a few drops of acetic acid.

5. The mixture is heated at 70°C and then it is allowed to stand at room temperature for overnight.
6. The mixture is centrifuged. The supernatant is discarded and the pellet is dissolved in sulphuric acid.
7. The extract is made upto a known volume and from this, a known volume of aliquot is taken in a conical flask.
8. A burette is filled with 0.01 *N* potassium permanganate solution (Reagent-D) and titration is done. The end-point is determined by appearance of pink colour, which should persist for some time.

Calculation

The amount of calcium present in the sample is calculated by the following formula.

1 ml of 0.01 N K MnO_4 solution is equivalent to 0.002 g of calcium.

Therefore, amount of calcium present in the sample =

$$\frac{d \times 0.002 \times b}{c \times a} \times 100$$

where,

a = Weight of sample.
b = Total volume of extract.
c = Volume of aliquot taken for estimation.
d = Average burette reading for sample.

6.22 Iron

Iron is a constituent of cytochrome and has an essential role in the synthesis of chlorophyll and many enzyme activities. It is absorbed by the plant from soil both in ferrous and ferric forms.

Principle

In estimating iron, it is at first brought to ferric form by the use of oxidizing agent, *i.e.*, potassium persulphate. After conversion, it is reacted with potassium thiocyanate to form red coloured ferric thiocyanate complex. The intensity of the colour appeared is then measured colorimetrically.

Reagent

(A) *Digestion mixture*: 20 ml of nitric acid (HNO_3) and 2 ml of sulphuric acid (H_2SO_4) are taken and 8 ml of 70 per cent perchloric acid ($HClO_4$) is added slowly to it. The mixture is carefully shaken to mix well.

(B) *Potassium persulphate* ($K_2S_2O_8$) *solution*: Saturated in distilled water (W/V).

(C) *Potassium thiocyanate* (KSCN) *solution*: 3 *N* in distilled water (W/V).

(D) *Standard iron solution*: 0.702 g ferrous ammonium sulphate [$FeSO_4(NH_4)_2SO_4, 6H_2O$] is dissolved in 100 ml of distilled water. 5 ml of concentrated H_2SO_4 and a few drops of potassium permanganate ($KMnO_4$) solution are added to this. The volume is made upto 1 litre with distilled water and the concentration of this solution will b 100 µg/ml.

Procedure

1. The preparation of ash and digestion of it is done by the same procedure as is done in estimation of phosphorus (Ref. 6.19. Procedure → Steps 1-2).
2. The digested material is made upto a known volume with distilled water.
3. A number of aliquots (0.5-1.0 ml) are taken separately into test-tubes. Each should be made upto 1 ml with distilled water.
4. 5 ml of concentrated sulphuric acid is added to each test-tube.
5. Then, 1 ml of saturated potassium persulphate solution (Reagent-B) and 2 ml of 3 *N* potassium thiocyanate solution (Reagent-C) are added to each test-tube and allowed to stand for 10 minutes for development of colour.
6. The red colouration will appear due to reaction with iron that is present in the sample. The Optical Density (O.D.) value of the coloured solution is then measured through 440 nm wavelength in a colorimeter against blank.
7. *Preparation of standard curve*: From the standard iron solution (100 µg/ml) (Reagent-D), different concentration gradients are made. To each of the solution, red colour is developed by the same procedure. The O.D. value of it is then measured through 440 nm wavelength in a colorimeter against blank. A standard curve is prepared by plotting the known concentrations of iron solution against respective O.D. value of each.

Calculation

From the standard curve, the amount of iron actually present in the sample is determined.

6.23 Chlorophyll

Chlorophylls are metallo-porphyrin pigments formed by the chelation of porphyrins with magnesium. These are the essential pigments in photosynthetic reaction in plants. There are several forms of chlorophyll pigments, viz., chlorophyll *a,b,c,d*, bacterio-chlorophyll etc. However, chlorophyll *a* is the most potent photosynthetic pigment in all photosynthetically active plants.

Principle

The pigment, chlorophyll is extracted with acetone and the intensity of the colour appeared is measured colorimetrically. From the readings, the amount of chlorophyll actually present in the sample is calculated.

Reagents

(A) *Acetone* (CH_3COCH_3): 80 per cent in distilled water (V/V).

Procedure

1. Known amount of tissue sample by weight is taken. This is extracted with 80 per cent chilled acetone (Reagent-A) and the extract is filtered or centrifuged (3000 g). The extraction procedure is repeated until the residue has no greenness in it.
2. The filtrate or supernatant is made upto a known volume with 80 per cent acetone. The Optical Density (O.D.) value of the coloured solution is then measured through 642.2 nm, 642.5 nm, and 660.0 nm wavelength in a colorimeter against blank.

Calculation

The chlorophyll concentration in the sample is calculated by the following formula.

$$\text{Total chlorophyll } (\mu g/ml) = 7.12\,(A_{660.0}) + 16.8\,(A_{642.5})$$
$$\text{Chlorophyll } a\ (\mu g/ml) = 9.93\,(A_{660.0}) - 0.777\,(A_{642.5})$$
$$\text{Chlorophyll } b\ (\mu g/ml) = 17.6\,(A_{642.2}) - 2.81\,(A_{660.0})$$

where,

A = Specific O.D. value at specific wavelength.

6.24 Total Anthocyanin

Anthocyanins are water-soluble, phenolic glycosides belonged to flavonoid pigments, having a basic structural unit, *i.e.*, C_{15} skeleton of flavone. The more common of such pigments are cyanidin-3-rutinoside, delphinidin, malvidin etc. Anthocyanins are important pigments determining colour of many fruits, *e.g.*, apple, plum, cherry, grapes, many berries etc. and thereby, criteria for consumers' acceptance.

Principle

Anthocyanin is extracted with ethanolic-hydrochloride and the intensity of the colour appeared is measured colorimetrically. From the reading, the amount of the pigment present is determined.

Reagents

(A) *Ethanol* (C_2H_5OH): 95 per cent in distilled water (V/V).

(B) *Hydrochloric acid* (HCl): 1.5 *N*.

(C) *Ethanol-hydrochloric acid mixture*: To be prepared by mixing 85 parts of Reagent A and 15 parts of Regent B by volume.

Procedure

1. Known amount of tissue sample by weight is taken. This is extracted with ethanol-hydrochloric acid mixture (Reagent-C). It is allowed to stand overnight at a temperaiure of about 4°C.
2. After extraction, the mixture is filtered through filter paper (Whatman No. 1). The extraction is repeated for maximum recovery of the pigment.
3. The extract is then made upto any known volume with the solvent.
4. A known volume of aliquot from the above extract is taken. It is filtered through fine (milli-pore) filter paper.
5. The filtered aliquot is maintained in darkness for about 2 hours with cover of the container.
6. This is then made upto a known volume with the solvent (for colour measurement). The Optical Density (O.D.) value of the solution is then measured through 535 nm wavelength in a colorimeter against blank.

Calculation

The amount of total anthocyanin in the sample is calculated as follows.

Total absorbance value for the sample (per 100g) =

$$\frac{e \times b \times c}{d \times a} \times 100$$

where,

a = Weight of sample.

b = Volume made for colour measurement.

c = Total volume made.

d = Volume of aliquot taken for estimation.

e = Specific O.D. value at 535 nm wavelength.

1 mg per millilitre of the solution is equivalent to the absorbance of 98.2.

Therefore, the amount of total anthocyanin present in the sample (mg/100g) =

$$\frac{\text{Total absorbance for the sample}}{98.2}$$

6.25 Ethylene Evolution

The olefin, ethylene has now been known to be perhaps the most potent compound in the regulation of physiological and biochemical process in plants. This *gas hormone* has been established to have a crucial role in the set and development of a vast majority of fruits and especially, in the control of their ripening, either in the pre-harvest or under post-harvest condition.

Principle

Gas-chromatography (GC) is by far, the most reliable method to determine the evolution of ethylene from plant tissue.

Procedure

1. Known amount of fruit sample is weighed and taken in a 100 ml conical flask containing a layer of glass-beads. The conical flask is tightly stoppered with rubber lid and sealed.

2. The sample in the conical flask is then incubated in an atmosphere of, 20°C ± 1°C and 16-hour illumination in a 24-hour cycle.
3. 1 ml of gas sample is taken out from the conical flask with a specially designed syringe and then it is injected into the GC set.
4. To make the standard, 10 μg of pure ethylene is injected into a 100 ml conical flask and this is kept in a similar atmospheric situation. 1 ml of the gas sample is taken out from the conical flask with the syringe and it is injected into the GC. The peak height of ethylene produced is determined.
5. From the standard peak height, the amount of ethylene gas is calculated and expressed as μl of the gas evolved per hour per Kg of sample.

6.26 Carbon Di-oxide Evolution

The evolution of a relatively high level of carbon di-oxide gas is recognized in many fruits at specific stage of their development *in situ*, or under post -harvest condition. This abrupt rise of the gas indicating higher respiratory activity of the tissue is known as *climacteric respiration* and signals the onset of ripening in them. Interestingly, this higher rate of respiratory activity is known to have stimulation from the hydrocarbon gas, ethylene, *i.e.*, produced.

Principle

The fruit tissues are to be made into pieces and straightway used for determination of the carbon di-oxide evolved in a suitable respiroscope.

Procedure

1. Known amount of fruit sample is cut into smaller pieces. Without delay, the weighed amount of pieces are placed in a suitable container. These are subjected to determination of the amount of carbon di-oxide evolved in a Warburg apparatus.
2. The carbon di-oxide gas evolved is measured manometrically at specific temperature (usually 30°C) and duration. It is expressed as ml of the gas evolved per hour per Kg of the sample.

7

Determination of Activity of Enzymes

7.1 Assay of Enzyme Activity

Enzymes are potent *organic catalysts* in bringing about various metabolic reactions in plant. They function over specific substrates and under a specific and narrow limits of temperature and *p*H. In many cases, activity of them also depends much on other adjuncts, *i.e.*, co-factors, available naturally or supplied from external source. Virtually, all enzymes are proteins with molecular weights ranging from 10 thousand to several millions and the amino acid residues per molecule of an enzyme may vary from one hundred to several thousand.

According to the type of reactions catalyzed, enzymes are conventionally grouped under the following 6 categories.

Table 11: Types of enzymes and their catalytic actions

Class	*Name*	*Type of reaction catalysed*	*Example*
1	Oxidoreductase	Transfer of electrons	Polyphenol oxidase
2	Transferase	Transfer of functional groups	Hexokinase
3	Hydrolase	Hydrolysis	Polygalacturonase
4	Lyase	Clevage of C-C, C-O, C-N bonds	Pyruvate decarboxylase
5	Isomerase	Transfer of groups within a molecule	Phosphogluco-isomerase
6	Ligase (Synthase)	Bond formation couple ATP hydrolysis	Citrate synthase

Each enzyme is uniquely labelled with a systematic number. For example, trypsin is identified as *EC 3.4.21.4*, where EC is abbreviated form of Enzyme Commission. In this code, the first number (3) denotes that the enzyme is in class 3, *i.e.*, a hydrolase, the second number (4) denotes that it is a protease which hydrolyzes peptide bonds, the third number (21) denotes that it is a serine protease while the fourth number (4) signifies the characteristics number of this enzyme.

Each enzyme has also a trival name which is an abbreviated term that is commonly used in its nomenclature. For example, the commonly used term polygalacturonase has its systematic name, *i.e.*, poly (1,4–α–D–galacturonide) glycanohydrolase and its number is *EC 3.2.1.15*.

The activity of an enzyme is determined in respect of its catalytic effect, *i.e.*, the rate of accumulation of the product, or the rate of consumption of the substrate under a given set of conditions.

The enzyme activity can be determined either by continuous (kinetic) or discontinuous (fixed time) methods. In the continuous method of assay, the change of absorbance in respect of the substrate or the product is measured at a specific wavelength, while the discontinuous method involves determination of the product formed by the enzyme after a fixed time interval.

Evidently, the control sets should always be maintained in determining activity of an enzyme and the conditions for the control should remain the same in all respects as with the test sample. The control is needed to be maintained to allow non-specific chemical reactions if any, which is not activated by the enzyme. A control (with inactivated enzyme) signifies an assessment of the extent of the non-enzymic reaction, whereas, a control (lacking substrate) evaluates any background reaction in the enzyme preparation.

The enzyme activity is expressed in terms of *units* and this is based upon the rate of the catalytic reaction. The term, *International Unit* (IU) refers to the amount of enzyme in a particular preparation which catalyzes the conversion of 1 micromole of a substrate to product in 1 minute under given conditions. The SI units of enzyme activity is the *Katal* (Kat) which represents the transformation of 1 mole of a substrate per second under a specified condition. The conversion factor between these units is stated below.

IU = Iμ mol min^{-1}

1 Kat = 6 x 10^7 U

1U = 1.7 x 10^{-8} Kat = 16.67 nano kat

The purity of an enzyme is expressed as *specific activity* which is the number of enzyme units per milligram of protein present in the solution. Thus, it may be expressed as follows.

$$\text{Specific activity (units/mg)} = \frac{\text{Total units of enzyme}}{\text{Total amount of protein}}$$

7.2 α–Amylase

α–amylase (1,4–α–D–glucan glucanohydrolase, *EC 3.2.1.1*) hydrolyzes the glycosidic bond α–1,4 which is found in a starch polymer. The increased activity of α–amylase is much apparent during germination of seeds, ripening of fruits, or where breakdown of starch is involved.

Principle

α–amylase activity is assayed by incubating the enzyme with the substrate (*i.e.*, starch) for a given period of time. The enzyme causes degradation of starch by hydrolysing 1,4 linkage, which is reacted with iodine to give rise to blue coloured complex. The intensity of colouration is then measured colorimetrically and compared with a standard curve. The decrease in intensity of colouration for a unit time is proportional to the activity of the enzyme.

Reagents

(A) *Tris–HCl buffer solution*: 0.01 *M*, *p*H 7.5.

(B) *Calcium chloride* ($CaCl_2$): 1 *M* in distilled water (W/V).

(C) *Potassium dihydrogen phosphate* (KH_2PO_4): 0.06 *M* in distilled water (W/V).

(D) *Starch* [$(C_6H_{10}O_5)n$] *solution*: 0.15 per cent in 0.06 *M* KH_2PO_4 (W/V).

(E) *Potassium iodide-iodine reagent*: 0.025 g iodine (I_2) and 0.4 g potassium iodide (KI) are dissolved in little amount of distilled water and the volume is made upto 100 ml with distilled water.

(F) *Standard starch solution:* 100 mg soluble starch [$(C_6H_{10}O_5)n$] is dissolved in little amount of distilled water and the volume is made upto 100 ml with distilled water. From this solution, 10 ml is taken and this is diluted to make 100 ml with distilled water. The concentration of this solution will be 0.1 mg/ml.

Procedure

1. *Extraction of enzyme*: Known amount of fresh tissue sample by weight is taken. This is homogenized with 0.01 *M* tris-HCl buffer (*p*H 7.5) (Reagent-A) containing 1 *M* calcium chloride (Reagent-B) for 3 hours at room temperature. The homogenate is centrifuged (10,000 g) at 4°C for 15 minutes and the supernatant is used as the enzyme extract for assay of the enzyme activity.
2. The substrate, i.e., 1 ml of 0.15% starch solution (Reagent-D) is taken in a test-tube. Then 0.5 ml enzyme extract is added to this and incubated at 30°C for 1 hour.
3. At the end of the incubation period, 1 ml potassium iodide-iodine reagent (Reagent-E) is added to this. The blue coloured complex is then appeared in the reaction mixture and the Optical Density (O.D.) value of the coloured solution is then measured through 620 nm wavelength in a colorimeter against blank.
4. An enzyme control is also run simultaneously with it. In preparation of it, an aliquot of enzyme extract is inactivated by heating in a boiling water-bath for 5 minutes and then it is added to the substrate solution.
5. *Preparation of standard curve*: From the standard starch solution (0.1 mg/ml) solution (Reagent-F), different concentration gradients are made. To each of the solution, blue colouration is developed by the same procedure. The O.D. value of it is then measured through 620 nm wavelength in a colorimeter against blank. A standard curve is prepared by plotting the known concentrations of glucose solution against respective O.D. value of each.

Calculation

From the standard curve, the amount of residual starch in the sample solution is determined. The enzyme activity is expressed as mg starch hydrolyzed per hour per gram of fresh tissue sample.

7.3 β-Amylase

β-amylase (1, 4–α–D glucan maltohydrolase, *EC 3.2.1.2*) hydrolyzes amylase and forms maltose units from it. Thus, this enzyme can hydrolyze the alternate bonds from the non-reducing end of the substrate. As with α–amylase, the increased activity of it is also observed during seed germination and fruit ripening.

Principle

β-amylase activity is assayed by incubating the enzyme with the substrate starch for a given period of time. The enzyme releases reducing sugars, which are reacted with dinitrosalicylic acid reagent to give rise to red coloured complex. The intensity of colouration is then measured colorimetrically and compared with a standard curve to determine the amount or released reducing sugars which is proportional to the activity of enzyme.

Reagents

(A) *Phosphate buffer solution*: 0.06 *M*, *p*H 7.0.

(B) *Sodium Chloride* (NaCl) *solution*: 0.5 *M* in distilled water (W/V).

(C) *Starch* [$(C_6H_{10}O_5)n$] *solution*: 0.2 per cent in 0.06 *M* phosphate buffer, *p*H 6.0 (W/V).

(D) *Ethylene diamine tetra acetic acid* ($C_{10}H_{16}N_2O_8$): 0.1 *M* in distilled water (W/V).

(E) *Dinitrosalicylic acid reagent*: 2.5 g 3, 5 dinitrosalicylic acid ($C_7H_4N_2O_7$) is dissolved in 50 ml of distilled water. 5 g sodium hydroxide (NaOH) and 50 g sodium potassium tartrate ($C_4H_4KNaO_6$, $4H_2O$) are seperately dissolved in 50 ml distilled water. Then dinitrosalicylic acid and sodium potassium tartrate solutions are mixed well and sodium hydroxide solution is added to this. The volume is made upto 250 ml with distilled water.

(F) *Standard maltose solution*: 100 mg maltose ($C_{12}H_{22}O_{11}$, H_2O) is dissolved in little amount of distilled water and the volume is made upto 100 ml with distilled water. From this solution, 10 ml is taken and this is diluted to make 100 ml with distilled water. The concentration of this solution will be 100 μg/ml.

Procedure

1. *Extraction of enzyme*: Known amount of fresh tissue sample by weight is taken. This is homogenized with 0.06 *M* phosphate buffer (*p*H 7.0) (Reagent-A) containing 0.5 *M* sodium chloride (Reagent-B) for 2 hours at room

temperature. The homogenate is centrifuged (20000 g) at 4°C for 15 minutes and the supernatant is used as the enzyme extract for assay of the enzyme activity.

2. The substrate mixture is prepared by mixing 1 ml of 0.2 per cent starch solution (Reagent-C) and 0.2 ml of 0.1 *M* ethylene diamine tetra acetic acid solution (Reagent-D). It is taken in a test-tube. Then 0.3 ml of enzyme extract is added to this and incubated at 30°C for 1 hour.
3. At the end of this incubation period, 3 ml of di-nitrosalicylic acid (Reagent-E) is added to this and then heated in a boiling water-bath for 10 minutes.
4. After cooling at room temperature, the Optical Density (O.D.) value of the coloured solution is then measured through 540 nm wavelength in a colorimeter against blank.
5. An enzyme control is also run simultaneously with it. In preparation of it, an aliquot of enzyme extract is inactivated by heating in a boiling water-bath for 5 minutes and then it is added to the substrate solution.
6. *Preparation of standard curve*: From the standard maltose solution (100 µg/ml) (Reagent-F), different concentration gradients are made. To each of the solution, red colouration is developed by the same procedure. The O.D. value of it is then measured through 540 nm wavelength in a colorimeter against blank. A standard curve is prepared by plotting the known concentrations of maltose solution against respective O.D. value of each.

Calculation

From the standard curve, the amount of released maltose in the sample solution is determined. The enzyme activity is expressed as µg maltose liberated per minute per gram of fresh tissue sample.

7.4 Pectin Methyl Esterase

Pectin methylesterase (Pectin pectyl-hydrolase, *EC 3.1.1.11*) is an enzyme protein of about 30000 molecular weight and an isoelectric point of about *p*H 9.0 as worked out in banana fruits. The activity of the enzyme is slightly increased during fruit ripening, causing an increased accumulation of the oxalate-soluble pectic fractions. However, it is not clearly related to the softening process for it acts in partial demethylation of the pectin, which forms as a substrate for activity of the enzyme, polygalacturonase. The activity of pectin methylesterase is known to be inhibited by sucrose or other sugar molecules.

Principle

The enzyme, pectin methylesterase brings about an increase of the carboxyl groups by partial demethylation of pectin. This increased level of carboxyl groups is determined titrimetrically against a standard alkali. The amount of alkali consumed for a given period of time is proportional to the activity of the enzyme.

Reagents

(A) *Sodium chloride* (NaCl) *solution*: 10 per cent in distilled water (W/V).

(B) *Ethylene diamine tetra acetic acid-sodium salt* ($C_{10}H_{14}N_2O_8Na_2$, $2H_2O$) *solution*: 1 per cent in distilled water (W/V).

(C) *Polyvinyl pyrrolidone (Mol. wt. 44000) solution*: 1 per cent in distilled water (W/V).

(D) *Sodium hydroxide* (NaOH) *solution*: 0.1 *N* in distilled water.

(E) *Sodium chloride* (NaCl) *solution*: 1 *M* in distilled water (W/V).

(F) *Pectin solution*: 1 per cent in 1 *M* NaCl solution (W/V).

Procedure

1. *Extraction of enzyme*: Known amount of fresh tissue sample by weight is taken. This is homogenized with a mixture of 10 per cent sodium chloride solution (Reagent-A), 1 per cent ethylene diamine tetra acid (sodium salt) (Reagent-B) and 1 per cent polyvinyl pyrrolidone (Reagent-C) at 4°C for 15 minutes. Polyvinyl pyrrolidone is added to bind the phenolic compounds which inhibit the enzyme activity. The homogenate is adjusted to *p*H 7.0 with 0.1 *N* sodium hydroxide solution (Reagent-D). Then it is centrifuged (30000 g) at 4°C for 10 minutes. The supernatant is used as the enzyme extract for assay of the enzyme activity.
2. The substrate mixture containing 20 ml of 1 per cent pectin solution (Reagent-F) and 5 ml of 1 *M* sodium chloride solution (Reagent-E) are taken in a beaker. Then, 10 ml of

distilled water is added to it and the beaker is placed over a water-bath at 30°C.

3. The reaction is started by the addition of 10 ml of enzyme extract with continuous stirring. The *p*H of the reaction mixture is adjusted to 7.5. The reaction is carried out at 30°C for 30 minutes.
4. At the end of the reaction period, 0.1 *N* sodium hydroxide solution (Reagent-D) is added to it, so that the *p*H of the reaction mixture is strictly maintained to 7.5. The amount of required 0.1 *N* sodium hydroxide solution to maintain the same *p*H is noted.
5. An enzyme control is also run simultaneously with it. In preparation of it, an aliquot of enzyme extract is inactivated by heating in a boiling water-bath for 5 minutes and then it is added to the substrate mixture.

Calculation

One milli-equivalent of carboxyl group is calculated by multiplying the amount of required sodium hydroxide solution with the used normality of it.

The enzyme activity is expressed as unit per gram of fresh tissue sample where unit refers to the amount of enzyme which releases one milli-equivalent of carboxyl group per minute per gram of fresh tissue sample.

7.5 Polygalacturonase

Polygalacturonase [Poly (1, 4–α–D-galacturonide) glycanohydrolase, *EC 3.2.1.15*] is an important pectolytic enzyme which is responsible for softening of fruits. Hydrolysis of the α–1, 4 linkage of polygalacturonates catalyzed by this enzyme is involved in the solubilization of cell-wall bound forms of the pectic polymers to convert to the water-soluble fractions. Most fruit tissues contain appreciable amount of this enzyme which exhibits a characteristic change in its activity, especially during ripening of fruits. The activity of this enzyme is also known to be highly associated with the endogenous level of ethylene in fruits.

Principle

Polygalacturonase activity is assayed by incubating the enzyme with the substrate, polygalacturonic acid for a given period of time. The enzyme releases reducing sugars, which are reacted with arsenomolybdate reagent in presence of alkaline copper tartrate solution to give rise to blue coloured complex. The intensity of colouration is then measured colorimetrically and compared with a standard curve to determine the amount of released reducing sugars which is proportional to activity of the enzyme.

Reagents

(A) *Sodium chloride* (NaCl) *solution*: 1 *M* in distilled water (W/V).

(B) *Sodium Chloride* (NaCl) *solution*: 0.15 *M* in distilled water (W/V).

(C) *Sodium acetate buffer solution*: 0.05 *M* *p*H 3.8.

(D) *Polygalacturonic acid-sodium salt solution*: 0.5 per cent in 0.05 *M* sodium acetate (Reagent-C) buffer solution, *p*H 3.8 (W/V).

(E) 2.5 g anhydrous sodium carbonate (Na_2CO_3), 2 g sodium bicarbonate ($NaHCO_3$), 2.5 g sodium potassium tartrate ($C_4H_4KNaO_6$, $4H_2O$) and 20 g anhydrous sodium sulphate (Na_2SO_4) are dissolved in distilled water and the volume is made upto 100 ml with distilled water.

(F) 15 g copper sulphate ($CuSO_4$, $5\,H_2O$) is dissolved in little amount of distilled water. One drop of concentrated sulphuric acid (H_2SO_4) is added to it and the volume is made upto 100 ml with distilled water.

(G) *Alkaline copper tartrate solution*: The reagent is freshly prepared as a proportion of 4 ml of Reagent-F and 96 ml of Reagent-E.

(H) *Arsenomolybdate reagent*: 25 g ammonium molybdate [$(NH_4)_6Mo_7O_{24}$, $4H_2O$] is dissolved in 450 ml of distilled water. Then 21 ml of concentrated sulphuric acid is added slowly to it and mixed well. 3 g disodium hydrogen arsenate (Na_2HAsO_4, $7H_2O$) is separately dissolved in 25 ml of distilled water. Solutions of ammonium molybdate and disodium hydrogen arsenate made above are mixed properly and stored at 37°C for 24 hours in dark.

(I) *Standard galacturonic acid solution*: 120.5 mg galacturonic acid monohydrate ($C_6H_{10}O_7$, H_2O) is weighed and taken in a volumetric flask. 5 ml of 0.05 *N* sodium hydroxide

(NaOH) solution is added to it and the volume is made upto 100 ml with distilled water. From this solution, 10 ml is taken and this is diluted to make 100 ml with distilled water. The concentration of this solution will be 100 µg/ml.

Procedure

1. *Extraction of enzyme*: Known amount of fresh tissue sample by weight is taken. This is homogenized with ice-cold distilled water for 1-2 minutes. The homogenate is centrifuged (2400 g) for 10 minutes at 4°C and the pellet is suspended in 1 *M* sodium chloride solution (Reagent-A) and adjusted to *p*H 6.0. The supernatant obtained after centrifugation is made to 75 per cent saturation (W/V) with solid ammonium sulphate $[(NH_4)_2 SO_4]$. The precipitated protein is taken out by centrifugation (10000 g) for 20 mintues. The pellet is suspended in 0.15 *M* sodium chloride solution (Reagent-B) and adjusted to *p*H 6.0. Then it is dialyzed overnight at 4°C against 0.15 *M* sodium chloride solution. After a brief centrifugation (12000 g), the supernatant is used as the enzyme extract for assay of the enzyme activity.
2. The substrate mixture is prepared by mixing 0.5 ml of 0.15 *M* sodium chloride solution (Reagent-B) and 1.5 ml of 0.5 per cent of polygalacturonic acid solution (Reagent-D). It is taken in a test-tube. Then 1 ml of enzyme extract is added to this and incubated at 30°C for 6 hours.
3. At the end of the incubation period, 1 ml of alkaline copper tartrate solution (Reagent-G) is added to this and then heated in a boiling water-bath for 10 minutes.
4. After cooling to room temperature, 1 ml of arsenomolybdate reagent (Reagent-H) is added to this to develop the blue colouration. The total volume of the reaction mixture is made upto 10 ml with distilled water. The Optical Density (O.D.) value of the coloured solution is then measured through 620 nm wavelength in a colorimeter against blank.
5. An enzyme control is also run simultaneously with it. In preparation of it, an aliquot of enzyme extract is inactivated by heating in a boiling water-bath for 5 minutes and then it is added to the substrate solution.

6. *Preparation of standard curve*: From the standard galacturonic acid solution (100 µg/ml) (Reagent-I), different concentration gradients are made. To each of the solution, blue coloration is developed by the same procedure. The O.D. value of it is then measured through 620 nm wavelength in a colorimeter against blank. A standard curve is prepared by plotting the known concentrations of galacturonic acid solution against respective O.D. value of each.

Calculation

From the standard curve, the amount of released galacturonic acid in the sample solution is determined. The enzyme activity is expressed as µ mol galacturonic acid liberated per minute per gram of fresh tissue sample.

7.6 Cellulase

Celulose (1,4–β–D–glucan, 4–glucanohydrolase, *EC 3.2.1.4*) is an important enzyme in regard degradation of cell-well and thus, leading to softening of ripening fruits. However, this enzyme from plant tissues is characterized by its ability to degrade carboxymethyl cellulose and incapability to degrade the native cellulose. The possible substrates of the enzyme are xyloglucan and β–1,4 linked hemicellulosic polysaccharide *in vivo* which are apparent in cell-walls of some fruits. There is evidence of increased activity of this enzyme during ripening of fruits.

Principle

Cellulase activity is assayed by incubating the enzyme with the substrate carboxymethyl cellulose for a given period of time. The enzyme releases reducing sugars, which are reacted with dinitrosalicylic acid reagent to give rise to red coloured complex. The intensity of colouration is then measured colorimetrically and compared with a standard curve to determine the amount of released reducing sugars which is proportional to the activity of enzyme.

Reagents

(A) *Phosphate buffer solution*: 0.05 *M*, *p*H 6.5.

(B) *Carboxymethyl cellulose solution*: 1 per cent in 0.05 *M* phosphate buffer, *p*H 6.5 (W/V).

(C) *Dinitrosalicylic acid reagent*: 2.5 g 3, 5 dinitrosalicylic acid ($C_7H_4N_2O_7$) is dissolved in 50 ml of distilled water. 5 g sodium hydroxide (NaOH) and 50 g sodium potassium tartrate ($C_4H_4KNaO_6$, $4H_2O$) are seperately dissolved in 50 ml distilled water. Then dinitrosalicylic acid and sodium potassium tartrate solutions are mixed well and sodium hydroxide solution is added to this. The volume is made upto 250 ml with distilled water.

(D) *Standard glucose solution*: 100 mg glucose ($C_6H_{12}O_6$) is dissolved in little amount of distilled water and the volume is made upto 100 ml with distilled water. From this solution, 10 ml is taken and this is diluted to make 100 ml with distilled water. The concentration of this solution will be 100 μg/ml.

Procedure

1. *Extraction of enzyme*: Known amount of fresh tissue sample by weight is taken. This is homogenized with chilled acetone ($CH_3CO\,CH_3$) and the defatted sample is prepared. This is extracted with 0.05 *M* phosphate buffer (*p*H 6.5) (Reagent-A) for 2 hours at 4°C. The homogenate is centrifuged (10000 g) at 4°C for 30 minutes and the supernatant is used as the enzyme extract for assay of the enzyme activity.
2. The substrate mixture is prepared by mixing 1 ml of 0.05 *M* phosphate buffer (*p*H 6.5) (Reagent-A) and 1 ml of 1 per cent carboxymethyl cellulose solution (Reagent-B). It is taken in a test-tube. Then 1 ml enzyme extract is added to this and incubated at 30°C for 6 hours.
3. At the end of the incubation period, 3 ml of dinitrosalicylic acid (Reagent-C) is added to it and then heated in a boiling water-bath for 10 minutes.
4. After cooling to room temperature, the Optical Density (O.D.) value of the coloured solution is then measured through 540 nm wavelength in a colorimeter against blank.
5. An enzyme control is also run simultaneously with it. In preparation of it, an aliquot of enzyme extract is inactivated by heating in a boiling water-bath for 5 minutes and then it is added to the substrate solution.

6. *Preparation of standard curve*: From the standard glucose solution (100 µg/ml) (Reagent-D), different concentration gradients are made. To each of the solution, red colouration is developed by the same procedure. The O.D. value of it and then measured through 540 nm wavelength in a colorimeter against blank. A standard curve is prepared by plotting the known concentrations of glucose solution against respective O.D. value of each.

Calculation

From the standard curve, the amount of glucose liberated in the sample solution is determined. The enzyme activity is expressed as µl mol glucose liberated per minute per gram of fresh tissue sample.

7.7 Invertase

Invertase (β–D–fructofuranoside fructohydrolase, *EC 3.2.1.26*) acts in hydrolyzing a sucrose molecule to form one molecule each of glucose and fructose. This enzyme accounts for the level of sucrose in fruits.

Principle

Invertase activity is assayed by incubating the enzyme with the substrate sucrose for a given period of time. The enzyme releases reducing sugars, which are reacted with aresnomolybdate reagent in presence of alkaline copper tartrate solution to give rise to blue coloured complex. The intensity of colouration is then measured colorimetrically and compared with a standard curve to determine the amount of released reducing sugars which is proportional to the activity of the enzyme.

Reagents

(A) *Sodium acetate buffer solution*: 0.2 *M*, *p*H 4.8.

(B) *Sucrose solution*: 0.4 *M* in 0.2 *M* sodium acetate buffer, *p*H 4.8 (W/V).

(C) 2.5 g anhydrous sodium carbonate (Na_2CO_3), 2 g sodium bicarbonate ($NaHCO_2$), 2.5 g sodium potassium tartrate (C_4H_4 $KNaO_6$, $4H_2O$) and 20 g anhydrous sodium sulphate (Na_2SO_4) are dissolved in distilled water and the volume is made upto 100 ml with distilled water.

(D) 15 g copper sulphate ($CuSO_4$, $5H_2O$) is dissolved in little amount of distilled water. One drop of concentration sulphuric acid (H_2SO_4) is added to it and the volume is made upto 100 ml with distilled water.

(E) *Alkaline copper tartrate solution*: The reagent is freshly prepared as a proportion of 4 ml of Reagent-D and 96 ml of Reagent-C.

(F) *Arsenomolybdate reagent*: 25 g ammonium molybdate [$(NH_4)_6$ Mo_7O_2, $4H_2O$] is dissolved in 450 ml of distilled water. Then 21 ml of concentrated sulphuric acid is added slowly to it and mixed well. 3 g disodium hydrogen arsenate ($Na_2HA_SO_4$, $7H_2O$) is dissolved in 25 ml of distilled water. Solutions of ammonium molybdate and disodium hydrogen arsenate made above are mixed properly and stored at 37° C for 24 hours in dark.

(G) *Standard glucose solution*: 100 mg glucose ($C_6H_{12}O_6$) is dissolved in little amount of distilled water and the volume is made upto 100 ml with distilled water. From this solution, 10 ml is taken and this is diluted to make 100 ml with distilled water. The conc. of this solution will be 100 µg/ml.

Procedure

1. *Extraction of enzyme*: Known amount of fresh tissue sample by weight is taken. This is extracted with 0.2 *M* sodium acetate buffer solution (*p*H 4.8) (Reagent-A) for 2 hours at 4°C. The homogenate is centrifuged (10000 g) at 4°C for 30 minutes and the supernatant is used as the enzyme extract for assay of the enzyme activity.
2. The substrate mixture is prepared by mixing 0.65 ml of 0.2 *M* sodium acetate buffer (*p*H 4.8) (Reagent-A) and 0.25 ml of 0.4 *M* sucrose solution (Reagent-B). It is taken in a test-tube. Then 0.1 ml of enzyme extract is added to this and incubated at 37°C for 1 hour.
3. At the end of the incubation period, 1 ml of alkaline copper tartrate solution (Reagent-E) is added to this and then heated in a boiling water-bath for 10 minutes.
4. After cooling to room temperature, 1 ml of arsenomolybdate reagent (Reagent-F) is added to this to develop blue colouration. The total volume of the reaction

mixture is made upto 10 ml with distilled water. The Optical Density (O.D.) value of the coloured solution is then measured through 620 nm wavelength in a colorimeter against blank.

5. An enzyme control is also run simultaneously with it. In preparation of it, an aliquot of enzyme extract is inactivated by heating in a boiling water-bath for 5 minutes and then it is added to the substrate solution.
6. *Preparation of standard curve*: From the standard glucose solution (100 μg/ml) (Reagent-G), different concentration gradients are made. To each of the solution, blue colouration is developed by the same procedure. The O.D. value of it is then measured through 620 nm wavelength in a colorimeter against blank. A standard curve is prepared by plotting the known concentrations of glucose solution against respective O.D. value of each.

Calculation

From the standard curve, the amount of released glucose in the sample solution is determined. The enzyme activity is expressed as μg glucose liberated per hour per milligram of fresh tissue sample.

7.8 β-galactosidase

β-galactosidase (β–D–galactoside galactohydrolase, *EC 3.2.1.23*) is considered a specific enzyme for ripening and is responsible for solubilization of galactose taking place during fruit ripening. Out of 3 isozymic forms of it, one (β–galactosidase II) is capable of degrading β–1, 4 linked galactan, which plays part in galactose solubilization from a galactan substrate. The increase in β–galactosidase is associated with increased solubility and depolymerization of the cell-wall pectin.

Principle

β–galactosidase activity is assayed by incubating the enzyme with the nitrophenyl substrate for a given period of time. The enzyme releases free- nitrophenol which are measured spectrophotometrically and compared, with a standard curve to determine the amount of it, which is proportional to the activity of the enzyme.

Reagents

(A) *Sodium acetate buffer solution*: 0.05 *M*, *p*H 6.0.

(B) *Sodium chloride* (NaCl) *solution*: 1 *M* in distilled water (W/V).

(C) *Sodium–EDTA solution*: 0.005 *M* in distilled water (W/V).

(D) *Sodium acetate buffer solution*: 0.05 *M*, *p*H 4.5.

(E) *p–nitrophenyl–β–D–galactopyranoside* ($C_{12}H_{15}NO_8$) *solution*: 0.025 *M* in 0.05 *M* sodium acetate buffer solution, *p*H 4.5 (W/V).

(F) *Sodium carbonate* (Na_2CO_3) *solution*: 1 *M* in distilled water (W/V).

(G) *Standard p-nitrophenol solution*: 100 mg *p*-nitrophenol ($C_6H_5NO_3$) is dissolved in little amount of distilled water and the volume is made upto 100 ml with distilled water. From this solution, 10 ml is taken and this is diluted to make 100 ml with distilled water. The concentration of this solution will be 100 μg/ml.

Procedure

1. *Extraction of enzyme*: Known amount of fresh tissue sample by weight is taken. This is homogenized with 0.05 *M* sodium acetate buffer solution (*p*H 6.0) (Reagent-A) containing 1 *M* sodium chloride (Reagent-B) at 4°C for 1 hour. The supernatant is obtained after centrifugation (20000 g) at 4°C for 25 minutes. Proteins are precipitated by the addition of solid ammonium sulphate [$(NH_4)_2SO_4$] to 80 per cent saturation (4°C) containing 0.005 *M* Na-EDTA salt (Reagent-C). The precipitated protein is recovered by centrifugation (20000 g) for 20 minutes. The pellet is dissolved in 0.05 *M* sodium acetate buffer solution (*p*H 6.0) at 4°C. After centrifugation (20000 g) for 10 minutes, the supernatant is dialyzed against 0.05 *M* sodium acetate buffer solution (*p*H 6.0) at 4°C for 24 hours. After centrifugation (12000 g), the supernatant is used as the enzyme extract for assay of the enzyme activity.
2. The substrate mixture is prepared by 0.8 ml of 0.05 *M* sodium acetate buffer solution (*p*H 4.5) (Reagent-D) and 0.1 ml of 0.025 *M p*–nitrophenol–β–D– galactopyranoside solution (Reagent-E). Then, 0.1 ml (5 to 25 μg of protein) of enzyme extract is added to this and incubated at 37°C for 1 hour.
3. At the end of the incubation period, the reaction is terminated by adding 2.0 ml of 1 *M* sodium carbonate solution (Reagent-F). The Optical Density (O.D.) value of

the free- nitrophenol in the solution is measured through 400 nm wavelength in a spectrophotometer against blank.

4. An enzyme control is also run simultaneously with it. In preparation of it, an aliquot of enzyme extract is inactivated by heating in a boiling water-bath for 5 minutes and then it is added to the substrate solution.
5. *Preparation of standard curve*: From the standard *p*-nitrophenol solution (100 µg/ml) (Reagent-G), different concentration gradients are made. The O.D. value of it is then measured through 400 nm wavelength in a spectrophotometer against blank. A standard curve is prepared by plotting the known concentrations of *p*-nitrophenol solution against respective O.D. value of each.

Calculation

From the standard curve, the amount of released *p*-nitrophenol in the sample solution is determined. The enzyme activity is expressed as µg *p*-nitrophenol released, per minute per gram of fresh tissue sample.

7.9 Protease

All protein-splitting enzymes are collectively known as proteases. Some protease enzymes are endopeptidase, while others cleave on the amino or carboxyl side of the amino acid residues. Proteases which are endopeptidase hydrolyze proteins into peptide fragments, whereas carboxy or aminopeptidases hydrolyze proteins into simple amino acids. Higher protease activity is noticeable in some fruits in particular. Papaya is a classic example in this group, in which it is abundantly found in the latex (as *papain, chymopapain, papaya peptidase* A) of young fruits, which ceases to flow when they undergo ripening. Among other fruits, the protease activity is also found in fig (*ficin*) and fruit and stem of pine-apple (*bromelain*) etc.

Principle

Protease activity is assayed by incubating the enzyme with the substrate protein for a given period of time. The enzyme releases amino acids which are reacted with the Folin & Ciocalteu's phenol reagent to give rise to blue coloured complex. The intensity of the colouration is then measured colorimetrically and compared with a standard curve to determine the amount of released amino acids which is proportional to the activity of the enzyme.

Reagents

(A) *Cysteine–EDTA buffer solution*: 0.71 g di-sodium hydrogen phosphate (Na_2HPO_4), 0.54 g cysteine hydrochloride ($C_3H_7NO_2S$, HCl) and 1.4 g ethylene diamine tetra acetic acid ($C_{10}H_{16}N_2O_8$) are dissolved in 80 ml of distilled water. The *p*H of the solution is then adjusted to 6.0 and the volume is made upto 100 ml with distilled water.

(B) *Caseine substrate solution*: 1 per cent in 0.05 *M* Na_2HPO_4 buffer solution, *p*H 6.0.

(C) *Trichloroacetic acid* ($C_2HCl_3O_2$): 10 per cent in distilled water (W/V).

(D) *Copper sulphate* ($CuSO_4$, $5H_2O$) *solution*: 0.5 per cent in distilled water (W/V).

(E) *Sodium potassium tartrate* ($KNaC_4H_4O_6$, $4H_2O$) *solution*: 1 per cent in distilled water (W/V).

(F) *Sodium carbonate* (Na_2CO_3) *solution*: 1 per cent in 1 *N* sodium hydroxide (NaOH) solution.

(G) *Protein reagent*: The reagent is freshly prepared as a proportion of 5 ml, 5 ml and 250 ml of Reagents-D, E and F respectively.

(H) *Folin and Ciocalteu's pheonl reagent*: 1 *N*.

(I) *Standard tyrosine solution*: 100 mg L- tyrosine ($C_9H_{11}NO_3$) is dissolved in little amount of distilled water and the volume is made upto 100 ml with distilled water. From this solution, 10 ml is taken and this is diluted to make 100 ml with distilled water. The concentration of this solution will be 100 µg/ml.

Procedure

1. *Extraction of enzyme*: Known amount of fresh tissue sample by weight is taken. This is homogenized with cysteine–EDTA buffer solution (*p*H 6.0) (Reagent-A) for 30 minutes. The homogenate is centrifuged (10000 g) for 15 minutes and the supernatant is used as the enzyme extract for assay of the enzyme activity.

2. The substrate mixture is prepared by mixing 2 ml of casein substrate solution (Reagent-B) and 0.5 ml of cysteine -EDTA- buffer solution (*p*H 6.0) (Reagent-A). It is taken in a test-tube. Then 1 ml enzyme extract is added to this and incubated at 40°C for 1 hour.
3. At the end of the incubation period, 2 ml of 10 per cent trichloroacetic acid solution (Reagent-C) is added and centrifuged (2000 g) at 4°C for 10 minutes. The supernatant is collected for estimation of enzyme hydrolyzed products.
4. An aliquot (1 ml) is taken in a test-tube and 4 ml of protein reagent (Reagent-G) is added to this. It is then incubated at room temperature for 10 minutes.
5. 0.5 ml of Folin & Ciocalteu's phenol reagent (Reagent-H) is added to the test-tube. The blue colouration will appear in the solution. Allowing for 10 minutes, the Optical Density (O.D.) value of the coloured solution is measured through 750 nm wavelength against blank.
6. An enzyme control is also run simultaneously with it. In preparation of it, an aliquot of enzyme extract is inactivated by heating in a boiling water-bath for 5 minutes and then it is added to the substrate solution.
7. *Preparation of standard curve*: From the standard tyrosine solution (100 µg/ml) (Reagent-I), different concentration gradients are made. To each of the solution, blue coloration is developed by the same procedure. The O.D. value of it is then measured through 750 nm wavelength in a colorimeter against blank. A standard curve is prepared by plotting the known concentrations of tyrosine solution against respective O.D. value of each.

Calculation

From the standard curve, the amount of released tyrosine in the sample solution is determined. The enzyme activity is expressed as µg tyrosine released per minute per gram of fresh tissue sample.

7.10 Lipase

The enzyme, lipase (Triacylglycerol acylhydrolase, *EC 3.1.1.3*) is a hydrolyzing one which brings about hydrolysis of a triglyceride molecule with ultimate formation of free- fatty acids and glycerol.

Principle

The enzyme, lipase releases free- fatty acids by hydrolysis of triglycerides. The quantity of fatty acids is measured titrimetrically by a standard alkali solution. The amount of alkali consumed for a given period of time is proportional to the activity of the enzyme.

Reagents

(A) *Acetone* (CH_3COCH_3).

(B) *Acetone-ether mixture*: 1 part of acetone is mixed with 1 part of di-ethyl ether (C_2H_5-O-C_2H_5) and thoroughly homogenized.

(C) *TRIZMA buffer solution*: 0.2 *M*, *p*H 8.0.

(D) *Lipase substrate*: (purified form should be obtained).

(E) *Thymolphthalein solution*: 0.1 per cent in distilled water (W/V).

(F) *Sodium hydroxide (NaOH) solution*: 0.1 *N* in distilled water.

Procedure

1. *Extraction of enzyme*: Known amount of fresh tissue sample by weight is taken. This is homogenized with chilled acetone (CH_3COCH_3) (Reagent-A). The homogenate is filtered and the residue is successively washed with acetone -ether mixture (Reagent-B) and ether. The residue is then extracted with 0.2 *M* TRIZMA buffer (*p*H 8.0) (Reagent-C) and centrifuged (15,000 g) for 10 minutes. The supernatant is used as the enzyme extract for assay of the enzyme activity.
2. The substrate mixture is prepared by mixing 10 ml of lipase substrate (Reagent-D) and 2 ml of 0.2 *M* TRIZMA buffer solution (*p*H 8.0) (Reagent-C). It is taken in a beaker and placed over a water-bath at 30°C.
3. The reaction is started by adding 3 ml of enzyme extract with thorough stirring. A few drops of thymolphthalein indicator solution (Reagent-E) is also added to it and the reaction is carried out at 30°C for 30 minutes.
4. At the end of the reaction period, 0.1 *N* sodium hydroxide solution (Reagent-F) is slowly added to the reaction mixture till the thymolphthaelin end-point is determined. The amount of 0.1 *N* sodium hydroxide solution required to reach the end-point is noted.

5. An enzyme control is run simultaneously with it. In preparation of it, an aliquot of enzyme extract is inactivated by heating it in a boiling water-bath for 5 minutes and then it is added to the substrate mixture.

Calculation

One milli-equivalent of free- fatty acid is calculated by multiplying the amount of required sodium hydroxide solution with the used normality of it.

The enzyme activity is expressed as unit per gram of fresh tissue, where unit refers to the amount of enzyme which releases one milli-equivalent free- fatty acid per minute per gram of fresh tissue sample.

7.11 Ascorbic Acid Oxidase

Ascorbic acid oxidase (L–ascorbate: Oxygen oxidoreductase, *EC 1.10.3.3*) is a copper containing enzyme having a crucial role in oxidation of ascorbic acid and regulating the levels of oxidized and reduced forms of glutathione.

Principle

Ascorbic acid oxidase activity is assayed by incubating the enzyme with the ascorbic acid substrate. The enzyme oxidizes ascorbic acid and the rate of degradation of it due to oxidation is determined spectrophotometrically. The change in Optical Density (O.D.) value for a unit time is proportional to the activity of the enzyme.

Reagents

(A) *Phosphate buffer solution*: 0.1 *M* *p*H 6.5.

(B) *Phosphate buffer solution*: 0.2 *M*, *p*H 6.2.

(C) *Ascorbic acid* ($C_6H_8O_6$) *solution*: 0.001 *M* in 0.2 *M* phosphate buffer, *p*H 6.2 (W/V).

Procedure

1. *Extraction of enzyme*: Known amount of fresh tissue sample by weight is taken. This is homogenized with 0.1 *M* phosphate buffer (*p*H 6.5) (Reagent-A) at 4°C. The homogenate is centrifuged (20000 g) at 4°C for 15 minutes and the supernatant is used as the enzyme extract for assay of the enzyme activity.

2. The substrate mixture is prepared by mixing 1.0 ml of 0.2 *M* phosphate buffer (*p*H 6.2) (Reagent-B) and 0.3 ml of 0.001 *M* ascorbic acid solution (Reagent-C). It is taken in quartz cuvette and inserted into the spectrophotometer, adjusting to zero absorbance. Then 0.2 ml of enzyme extract and 1.5 ml of distilled water are added to it and incubated at 30°C.
3. The change in Optical Density (O.D.) value is measured through 265 nm wavelength in a sepctrophotometer from the start and continued upto 5 minutes with interval of 30 seconds.

Calculation

A standard curve is prepared by plotting the O.D. values against time. From the linear phase of reaction, the change in O.D. value per minute is calculated and the enzyme activity is expressed as change in O.D. value per minute per gram of fresh tissue sample.

7.12 Polyphenol Oxidase

Polyphenol oxidase (Monophenol dihydroxyphenylalanine: Oxygen oxidoreductase, *EC 1.14.18.1*) oxidises the phenolic compounds, *e.g.*, chlorogenic acid followed by a free radical polymerization causing formation of quinones, responsible for darkening of freshly cut tissues. Polyphenol oxidase includes catechol oxidase and laccase. Fresh fruits are appreciable sources of this enzyme. The enzyme is known to play part against pathogenic attack of the plant parts.

Principle

Polyphenol oxidase activity is assayed by incubating the enzyme with the substrate, catechol. The enzyme oxidizes catechol to form quinone and the rate of formation of it due to oxidation is determined spectrophotometrically. The change in Optical Density (O.D.) value for a unit time is proportional to the activity of the enzyme.

Reagents

(A) *Tris-HCl buffer solution*: 0.05 *M*, *p*H 7.2.

(B) *Sorbitol* ($C_6H_{14}O_6$) *solution*: 0.4 *M* in distilled water (W/V).

(C) *Sodium chloride* (NaCl) *solution*: 0.01 *M* in distilled water (W/V).

(D) *Phosphate buffer solution*: 0.1 *M* *p*H 6.5.

(E) *Catechol* ($C_6H_6O_3$) *solution*: 0.01 *M* in 0.1 M phosphate buffer, *p*H 6.5 (W/V).

Procedure

1. *Extraction of enzyme*: Known amount of fresh tissue sample by weight is taken. This is homogenized with 0.05 *M* tris-HCl buffer solution (*p*H 7.2) (Reagent-A) containing 0.4 *M* sorbitol ($C_6H_{14}O_6$) (Reagent-B) and 0.01 M sodium chloride (NaCl) (Reagent-C) at 2°C for 15 minutes. The homogenate is centrifuged (20000 g) at 2°C for 1.5 minutes and the supernatant is used as the enzyme extract for assay of the enzyme activity.
2. The substrate mixture is prepared by mixing 2.5 ml of 0.1 *M* phosphate buffer (*p*H 6.5) (Reagent-D) and 0.3 ml of 0.01 M-catechol solution (Reagent-E). It is taken in a quartz cuvette and inserted into the spectrophotometer, adjusting to zero absorbance. Then 0.2 ml of enzyme extract is added to it and incubated at 30°C.
3. The change in Optical Density (O.D.) value is measured through 495 nm wavelength in a spectrophotometer from the start and continued upto 5 minutes with interval of 30 seconds.

Calculation

A standard curve is prepared by plotting the O.D. values against time. From the linear phase of reaction, the change in O.D. value per minute is calculated and the enzyme activity is expressed as change in O.D. value per minute per gram of fresh tissue sample.

7.13 Peroxidase

Peroxidase (Donor: Hydrogen peroxide oxidoreductase, *EC 1.11.1.7*) is an iron containing enzyme. Plant peroxidase activity is apparent in several isozymic forms.

Principle

Peroxidase activity is assayed by incubating the enzyme with the substrate, *o*-dianisidine. The enzyme oxidizes the substrate and the rate of degradation of it due to oxidation is determined spectrophotometrically. The change in Optical Density (O.D.) value for a unit time is proportional to the activity of the enzyme.

Reagents

(A) *Phosphate buffer solution*: 0.1 *M*, pH 6.5 .

(B) *o-dianisidine* ($C_{14}H_{16}N_2O_2$) *solution*: 0.1 per cent in methanol (W/V).

(C) *Hydrogen peroxide* (H_2O_2) *solution*: 0.2 *M*.

Procedure

1. *Extraction of enzyme*: Known amount of fresh tissue sample by weight is taken. This is homogenized with 0.1 *M* phosphate buffer (*p*H 6.5) (Reagent-A) at 4°C. The homogenate is centrifuged (18000 g) at 4°C for 15 minutes and the supernatant is used as the enzyme extract for assay of the enzyme activity.
2. The substrate mixture is prepared by mixing 3.5 ml of 0.1 *M* phosphate buffer (*p*H 6.5) (Reagent-A) and 0.3 ml of 0.1 per cent *o*-dianisidine solution (Reagent-B). It is taken in a quartz cuvette and inserted into the spectrophotometer, adjusting to zero absorbance. Then 0.2 ml of enzyme extract and 0.2 ml of 0.2 *M* hydrogen peroxide solution (Reagent-C) are added to it and incubated at 30°C.
3. The change in Optical Density (O.D.) value is measured through 430 nm wavelength in a spectrophotometer from the start and continued upto 5 minutes with interval of 30 seconds.

Calculation

A standard curve is prepared by plotting the O.D. values against time. From the linear phase of reaction, the change in O.D. value per minute is calculated and the enzyme activity is expressed as change in O.D. value per minute, per gram of fresh tissue sample.

APPENDICES

APPENDIX I

Botanical Names of Fruits Referred to in the Text

Common name	*Botanical name*	*Family*
Almond	*Prunus dulcis* (Mill.) D.A. Webb.	Rosaceae
Aonla (Indian gooseberry)	*Phyllanthus emblica* L.	Euphorbiaceae
Apple	*Malus pumila* Mill.	Rosaceae
Apricot	*Prunus armeniaca* L. and related species	Rosaceae
Bael (Wood apple)	*Aegle marmelos* Correa	Rutaceae
Banana	*Musa paradisiaca* L. and related species	Musaceae
Ber (Indian plum)	*Ziziphus mauritiana* Lam., *Z jujube* L. and related species	Rhamnaceae
Bullock's heart	*Annona reticulata* L.	Annonaceae
Cape-gooseberry	*Physalis peruviana* L.	Solanaceae
Cashew	*Anacardium occidentale* L.	Anacardiaceae
Cherry	*Prunus avium* L., *P. cerasus* L. and related species	Rosaceae
Chestnut	*Castanea* species	Fagaceae
Custard apple	*Annona squamosa* L.	Annonaceae
Date	*Phoenix dactylifera* L.	Palmae
Elephant apple (Wood apple)	*Feronia limonia* (L.) Sw.	Rutaceae
Fig	*Ficus carica* L.	Moraceae
Grape	*Vitis vinifera* L. and related species	Vitaceae
Grapefruit	*Citrus paradisi* Macfad.	Rutaceae
Guava	*Psidium guajava* L.	Myrtaceae

Contd...............

Appendix I–Contd......

Common name	*Botanical name*	*Family*
Hazel-nut	*Corylus avellana* L.	Corylaceae
Jackfruit	*Artocarpus heterophyllus* Lamk.	Moraceae
Jamun (Black plum)	*Syzygium cuminii* Skeels	Myrtaceae
Karonda (Carissa)	*Carissa carandas* L.	Apocynaceae
Lemon	*Citrus limon* Burm., related species and hybrids	Rutaceae
Lime (Sour lime)	*Citrus aurantifolia* Swingle	Rutaceae
Litchi	*Litchi chinensis* Sonn.	Sapindaceae
Longan	*Euphoria longan* St.	Sapindaceae
Mandarin orange	*Citrus reticulata* Blanco	Rutaceae
Mango	*Mangifera indica* L.	Anacardiaceae
Nor (Star-gooseberry)	*Phyllanthus acidus* Sk.	Euphorbiaceae
Papaya	*Carica papaya* L.	Caricaceae
Peach	*Prunus persica* (L.) Bats. and related species	Rosaceae
Pear	*Pyrus communis* L. and related species	Rosaceae
Pecan-nut	*Carya illinoensis* (Wang.) K. Koch.	Juglandaceae
Phalsa (Grewia)	*Grewia subinaequalis* D.C.	Tiliaceae
Pine-apple	*Ananas comosus* Mer.	Bromeliaceae
Pistachio-nut	*Pistacia vera* L.	Anacardiaceae
Plum	*Prunus domestica* L., *P. salicina* Lindl. and related species	Rosaceae
Pomegranate	*Punica granatum* L.	Punicaceae
Sapota	*Achras zapota* L.	Sapotaceae
Shaddock	*Citrus grandis* Osb.	Rutaceae
Star-apple (Water-apple)	*Syzygium samarangense* (Blume) Mer. & Per.	Myrtaceae
Streculia-nut	*Streculia foetida* L.	Sterculiaceae
Strawberry	*Fragaria species* and hybrids	Rosaceae
Sweet orange	*Citrus sinensis* Osb.	Rutaceae
Walnut	*Juglans regia* L. and related species	Juglandaceae
Water-chestnut	*Trapa natans* L. var. *bispinosa* (Roxb.) and *T. natans* L. var. *quadrispinosa* (Roxb.)	Trapaceae
Water-melon	*Citrullus vulgaris* L.	Cucurbitaceae

APPENDIX II

Conversion Factors (Approximate)

Quantity	*Metric unit or Imperial unit*	*Imperial unit to Metric unit*
Length	1 centimetre (cm) = 0.394 inch (in)	1 inch (in) = 25.4 millimetre (mm)
	1 metre (m) = 3.28 foot (ft)	1 foot (ft) = 30.5 centimetre (cm)
	1 metre (m) = 1.09 yard (yd)	1 yard (yd) = 0.914 metre (m)
	1 kilometre (km) = 0.621 mile	1 mile = 1.61 kilometre (km)
Mass	1 gram (g) = 0. 0353 ounce (oz)	1 ounce (oz) = 28.3 gram (g)
	1 kilogram (kg) = 2.2 pound (lb)	1 pound (lb) = 454 gram (g)
	1 tonne (t) = 0.984 ton	1 ton = 1.02 tonne (t)
Area	1 square centimetre (cm^2) = 0.155 square inch (in^2)	1 square inch (in^2) = 6.45 square centimetre (cm^2)
	1 square metre (m^2) = 10.8 square foot (ft^2)	1 square foot (ft^2) = 929 square centimetre (cm^2)
	1 square metre (m^2) = 1.2 square yard (yd^2)	1 square yard (yd^2) = 0.836 square metre (m^2)
	1 hectare (ha) = 2.47 acre (ac)	1 acre (ac) = 0.405 hectare (ha)
	1 square kilometre (km^2) = 0.386 square mile	1 square mile = 2.59 square kilometre (km^2)

Contd.........

Appendix II–Contd.......

Quantity	*Metric unit or Imperial unit*	*Imperial unit to Metric unit*
Volume	1 cubic centimetre (cm^3) = 0.0610 cubic inch (in^3)	1 cubic inch (in^3) = 16.4 cubic centimetre (cm^3)
	1 cubic metre (m^3) = 35.3 cubic foot (ft^3)	1 cubic foot (ft^3) = 28.3 cubic decimetre (dm^3)
	1 cubic metre (m^3) = 1.31 cubic yard (yd^3)	1 cubic yard (yd^3) = 0.765 cubic metre (m^3)
	1 cubic metre (m^3) = 27.5 bushel (bus)	1 bushel (bus) = 0.0364 cubic metre (m^3)
	1 millilitre (ml) = 0.0352 fluid ounce (fl oz)	1 fluid ounce (fl oz) = 28.4 millilitre (ml)
	1 litre (l) = 1.76 pint (pt)	1 pint (pt) = 568 millilitre (ml)
	1 cubic metre (m^3) = 220 gallon (gal)	1 gallon (gal) = 4.55 litre (l)
Force	1 newton (N) = 0.225 pound force (lbf)	1 pound force (lbf) = 4.45 newton (N)

APPENDIX III

Proximate Principles of Some Fruits

Name of fruit	*Moisture (%)*	*Carbohydrate (%)*	*Protein (%)*	*Fat (%)*	*Fibre (%)*	*Minerals (%)*	*Calcium (mg/100g)*	*Phosphorus (mg/100g)*	*Iron (mg/100g)*	*Vitamin C (mg/100g)*
Aonla (*Phyllantus emblica*)	81.8	13.7	0.5	0.1	3.4	0.5	50	20	1.2	600
Apple (*Malus pumila*)	84.6	13.4	0.2	0.5	1.0	0.3	10	14	1.0	1.0
Apricot (*Prunus americana*)	85.3	11.6	1.0	0.3	1.1	0.7	20	25	2.2	6.0
Avocado (*Persea americana*)	73.6	0.8	1.7	22.8	–	1.1	10	80	0.7	–
Beal (*Aegle marmelos*)	61.5	31.8	1.8	0.3	2.9	1.7	85	50	0.6	8
Banana (*Musa paradisiaca*)	70.1	27.2	1.2	0.3	0.4	0.8	17	36	0.9	7
Ber (*Zizyphus* species)	81.6	17.0	0.8	0.3	–	0.3	4	9	1.8	76
Bilimbi (*Averrhoa bilimbi*)	94.4	3.5	0.5	0.3	1.0	0.3	15	10	1.2	32

Contd.......

Appendix III–Contd....

Name of fruit	*Moisture (%)*	*Carbohydrate (%)*	*Protein (%)*	*Fat (%)*	*Fibre (%)*	*Minerals (%)*	*Calcium (mg/100g)*	*Phosphorus (mg/100g)*	*Iron (mg/100g)*	*Vitamin C (mg/100g)*
Blackberry (*Rubus fruitcosus*)	87.2	6.7	1.3	0.5	3.8	0.5	30	20	4.3	9
Bullock's heart (*Annona reticulata*)	76.8	15.7	1.4	0.2	5.2	0.7	10	10	0.6	5
Cape-goodeberry (*Physalis peruviana*)	82.9	11.1	1.8	0.2	3.2	0.8	10	67	2.0	49
Carambola (*Averrhoa carambola*)	91.9	6.1	0.7	0.1	0.8	0.4	4	11	–	–
Cashew apple (*Anacardium occidentale*)	86.3	12.3	0.2	0.1	0.9	0.2	10	10	0.2	180
Cherry (*Prunus avium*)	83.4	13.8	1.1	0.5	0.4	0.8	24	25	1.3	7
Cherimoyer (*Annona cherimolia*)	76.4	20.2	1.3	0.3	1.5	0.3	30	20	0.4	7
Custard apple (*Annona sauamosa*)	70.5	23.5	1.6	0.4	3.1	0.9	17	47	1.5	37
Date (*Phoenix dactylifera*)	59.2	33.8	1.2	0.4	3.7	1.7	22	38	–	–
Durian (*Durio zibethinus*)	58.0	34.1	2.8	3.9	–	1.2	10	50	1.0	–

Contd......

Appendix III–Contd......

Name of fruit	*Moisture (%)*	*Carbohydrate (%)*	*Protein (%)*	*Fat (%)*	*Fibre (%)*	*Minerals (%)*	*Calcium (mg/100g)*	*Phosphorus (mg/100g)*	*Iron (mg/100g)*	*Vitamin C (mg/100g)*
Elephant apple (*Limonia acidissima*)	64.2	18.1	7.1	3.7	5.0	1.9	130	110	0.6	3
Fig (*Ficus carica*)	88.1	7.6	1.2	0.2	2.2	0.6	80	30	1.0	5
Gaub (*Diospyros* species)	69.6	26.6	1.4	0.1	1.5	0.8	58	27	–	–
Grape (*Vilis vinifera*)	82.2	13.1	0.6	0.4	2.8	0.9	20	23	0.5	1
Grapefruit (*Citrus paradisi*)	88.5	10.0	1.0	0.1	–	0.4	30	30	0.2	1
Guava (*Psidium guajava*)	81.7	11.2	0.9	0.3	5.2	0.7	10	28	1.4	212
Hog plum (*Spondias pinnata*)	90.3	4.5	0.7	0.3	1.0	0.5	36	11	3.9	21
Jackfruit (*Artocarpus heterophyllus*)	76.2	19.8	1.9	0.1	1.1	0.9	20	41	0.5	7
Jamun (*Syzgium cuminii*)	83.7	14.0	0.7	0.3	0.9	0.4	15	15	1.2	18
Monkey jackfruit (*Artocarpus lakoocha*)	82.1	13.3	0.7	1.1	2.0	0.8	50	20	0.5	135

Contd.......

Appendix III–Contd......

Name of fruit	*Moisture (%)*	*Carbohydrate (%)*	*Protein (%)*	*Fat (%)*	*Fibre (%)*	*Minerals (%)*	***Calcium (mg/100g)***	***Phosphorus (mg/100g)***	***Iron (mg/100g)***	***Vitamin C (mg/100g)***
Lemon (*Citrus limon*)	85.0	11.1	1.0	0.9	1.7	0.3	**70**	**10**	**2.3**	39
Lichi (*Litchi chinensis*)	84.1	13.6	1.1	0.2	0.5	0.5	**10**	**35**	0.7	31
Loquat (*Eriobotrya japonica*)	88.2	9.6	0.6	0.3	0.8	0.5	**30**	20	1.3	–
Mango (*Mangifera indica*)	81.0	16.9	0.6	0.4	0.7	0.4	14	16	1.3	16
Mangosteen (*Garcinia mangostana*)	84.9	14.3	0.5	0.1	–	0.2	10	20	0.2	–
Mulberry (*Morus* species)	86.5	10.3	1.1	0.4	1.1	0.6	70	30	2.3	12
Musk-melon (*Cucumis melo*)	95.2	3.5	0.3	0.2	0.4	0.4	32	14	1.4	26
Palmyra (*Borassus flabellifer*)	77.2	20.7	0.7	0.2	0.5	0.7	9	33	–	–
Papaya (*Carica papaya*)	90.8	7.2	0.6	0.1	0.8	0.5	17	13	0.5	57
Passion fruit (*Passiflora edulis*)	76.3	12.4	0.9	0.1	9.6	0.7	10	60	2.0	25

Contd........

Appendix III–Contd......

Name of fruit	Moisture (%)	Carbohydrate (%)	Protein (%)	Fat (%)	Fibre (%)	Minerals (%)	Calcium (mg/100g)	Phosphorus (mg/100g)	Iron (mg/100g)	Vitamin C (mg/100g)
Peach (*Prunus persica*)	86.0	10.5	1.2	0.3	1.2	0.8	15	41	2.4	6
Pera (*Pyrus communis*)	86.0	11.9	0.6	0.2	1.0	0.3	8	15	0.5	–
Persimmon (*Diospyros kaki*)	80.0	17.9	0.7	0.2	0.9	0.3	15	10	0.3	33
Phalsa (*Grewia asiatica*)	80.8	14.7	1.3	0.9	1.2	1.1	129	39	3.1	22
Pine-apple (*Ananas comosus*)	87.8	10.8	0.4	0.1	0.5	0.4	20	9	1.2	39
Plum (*Prunus domestica*)	86.9	11.1	0.7	0.5	0.4	0.4	10	12	0.6	5
Pomegranate (*Punica granatum*)	78.0	14.5	1.6	0.1	5.1	0.7	10	70	0.3	16
Pummelo (*Citrus maxima*)	88.0	10.2	0.6	0.1	0.6	0.5	30	30	0.3	20
Quince (*Cydonia oblonga*)	85.7	11.9	0.3	0.1	1.7	0.3	10	20	0.4	11
Raspberry (*Rubus* species)	84.8	10.6	1.0	2.1	0.1	1.8	40	110	2.3	30

Contd.....

Appendix III–Contd......

Name of fruit	*Moisture (%)*	*Carbohydrate (%)*	*Protein (%)*	*Fat (%)*	*Fibre (%)*	*Minerals (%)*	*Calcium (mg/100g)*	*Phosphorus (mg/100g)*	*Iron (mg/100g)*	*Vitamin C (mg/100g)*
Rose-apple (*Syzygium jambos*)	89.1	8.5	0.7	0.2	1.2	0.3	10	30	0.5	3
Sapota (*Achras zapota*)	73.7	21.4	0.7	1.1	2.6	0.5	28	27	2.0	6
Star-apple (*Syzygium samarengense*)	91.9	6.3	0.3	0.2	1.0	0.3	5	10	–	–
Strawberry (*Fragaria* species)	87.8	9.8	0.7	0.2	1.1	0.4	30	30	1.8	52
Sour lime (Lime) (*Citrus aurantifolia*)	48.6	10.9	1.5	1.0	1.3	0.7	90	20	0.3	63
Sweet lime (*Citrus limettioides*)	90.5	7.3	0.7	0.3	0.7	0.5	30	20	0.7	45
Sweet orange (*Citrus sinensis*)	88.4	9.3	0.8	0.3	0.5	0.7	40	30	0.7	50
Tomatillo (*Physalis ixocarpa*)	91.7	5.8	0.7	0.6	0.6	0.6	7	40	1.4	2
Water-melon (*Citrullus vulgaris*)	95.8	3.3	0.2	0.2	0.2	0.3	11	12	7.9	1

Bibliography

Abeles, F.B., Morgan, P. W. and Saltveit, M.E. Jr. (1992). *Ethylene in Plant Biology*. Academic Press, Inc., New York, U.S.A.

Abu-Sarra, A.F. and Abu-Goukh, A.A. (1992). Changes in pectinesterase, polygalacturonase and cellulase activity during mango fruit ripening. *Journal of Horticultural Science*. 67 : 561-568.

Ali, S.L. and Mazumdar, B.C. (1994). Effects of applications of plant growth regulators to papaya on the output and proteolytic activity of latex tapped from immature fruits. *Journal of Horticultural Science*. 69 (5) : 805-807.

Anon. (1995). *Official Methods of Analyses of AOAC International* (Ed. by P. Gunniff). AOAC International, Suite 400, 2200 Wilson Boulevard, Virginia, U.S.A.

Arnon, D.I. (1959). Copper enzymes in isolated chloroplasts. Polyphenoloxidase in *Beta vulgaris*. *Plant Physiology*. 24: 1-15.

Banerji, A.K. (Professor, Dept. of Biochemistry, Calcutta University, Calcutta, India). Personal communication to the principal author (1996).

Barbier, M. and Thibault, J.F. (1982). Pectic substances of cherry fruits. *Phytochemistry*. 21 (1): 111-115.

Bhattacharyya, A.K., Roy, G.C. and Mazumdar, B.C. (1988). De-greening of carambola fruits by steeping with brine water. *Science and Culture*. 54 (11): 343-375.

Biocon India (P) Ltd., Bangalore, India. Personal communication to the principal author (1992).

Blumenkrantz, N. and Asboe-Hansen, G. (1973). New method for quantitative determination of uronic acids. *Analytical Biochemistry*. 54 : 484-489.

Bonner, J. and Varner, J.E. (1965). *Plant Biochemistry*. Academic Press, New York, U.S.A.

Bradford, M.M. (1976). A rapid and sensitive method for the quantitation of microgram quantities of protein utilising the principle of protein-dye binding. *Analytical Biochemistry*. 72: 248-254.

Brett, C.T. and Hillman. J.R. (1985). *Biochemistry of Plant Cell Walls*. Cambridge University Press, U.K.

Browning, D.R. (1969). *Chromatography*. McGraw-Hill, London, U.K.

Carre, M.H. and Haynes, D. (1922). Estimation of pectin as calcium pectate and application of this method to determine soluble pectin in apples. *Biochemical Journal*. 16: 60.

Chatterjee, (Mrs) A. (Retired, Khaira Professor of Chemistry, Calcutta University, Calcutta, India). Personal communication to the principal author (1993).

Chinoy, J.J. (Professor of Botany and Director, University School of Sciences, Gujarat University). Personal Communication to the principal author (1967).

Cruess, W.V. (1977). *Commercial Fruit and Vegetable Products*. Allied Scientific Publishers, Bikaner, India.

Datta, Mukul, Manjari, Mukhopadhyay, Paramita, Majumdar, K. and Mazumdar, B.C. (1998). A survey of agro-waste materials as sources of pectin: part II. *Applied Nutrition*. 23(1): 28-29.

Davis, K.M., Hobson, G.E. and Grierson, D. (1990). Silver ions inhibit the ethylene stimulated production of ripening related *m* RNAs in tomato fruits. *Journal of Plant Physiology*. 135 (6): 708-713.

Faust, Miklos (Fruit Laboratory, USDA Agricultural Research Service, Beltsville Agricultural Research Center, Beltsville, Maryland 20705, U.S.A.). Personal communication to the principal author (1992).

Fisher, J. and Arnold, J.R.P. (1999). *Instant Notes in Chemistry for Biologists*. Viva Books Private Limited, New Delhi, India.

Fisher, R.L. and Bennet, C.A. (1991). Role of cell-wall hydrolysis in fruit ripening. *Annual Review of Plant Physiology and Plant Molecular Biology*. 42: 675-703.

Goering, H.K. and Van Soet, P.J. (1970). *Fibre Analysis*. Agricultural Research Service (*Hand Book*). 379, 20-25.

Goodwin, T.W. and Mercer, E.J. (1988). *Introduction to Plant Biochemistry*. Pergamon Press, U.K.

Gopalan, C., Rama Sastri, B.V. and Balasubramaniam, S.C. (Revised and updated by Narasinga Rao, B.S., Deosthale, Y.G. and Pant, K.C.) (1993). *Nutritive Value of Indian Foods*. National Institute of Nutrition, Indian Council of Medical Research, Hyderabad, India.

Guha, Gopa and Mazumdar, B.C. (1996). Vitamin C content in different layers of fruits. *Applied Nutrition*. 21 (2): 40 .

Hayes, W.B. (1970). *Fruit Growing in India*. Kitabistan, Allahabad, India.

Heldt, H.W. (1997). *Plant Biochemistry and Molecular Biology*. Oxford University Press, Oxford, U.K.

Hobson, G.E. (1962). Determination of polygalacturonase in fruits. *Nature*. U.K., 195 :805.

Hobson, G.E. (1963). Pectin esterase in normal and abnormal tomato fruit. *Biochemical Journal*. 86: 358-365.

Jayaraman, J. (1981). *Laboratory Manual in Biochemistry*. Wiley Eastern Limited, New Delhi, India.

Kalra, Y. P. (1988) *Handbook of Reference Methods for Plant Analysis*. Soil and Plant Analysis Council, Inc. CRC Press, New York, U.S.A.

Kertesz, Z.I. (1951). *The Pectic Substances*. Interscience Publishers, Inc., New York, U.S.A.

Kumri, M., Majumder, K. and Mazumdar, B.C. (1999). Comparative study of healthy and malformed shoots and inflorescences of mango in respect of some biochemical components. *Science and Culture*. 65 (3 and 4): 105-106.

Lal, G., Siddapa, G.S. and Tandon, G.L. (1986). *Preservation of Fruits and Vegetables*. Indian Council of Agricultural Research, New Delhi, India.

Lane, S.H. and Eynon, L. (1923). Determination of reducing sugar by Fehling's solution with methylene blue as an indicator. *Journal of Society of Chemical Industry*, 42: 32.

Leeshem, Y.L., Halevy, A.H. and Frunkel, C. (1986). *Process and Control of Plant Senescence*. Elsevier Science Publishing Co., Inc., U.S.A.

Leopold, A.C. (Professor of Horticulture, Purdue University, Lafayette, Indiana, U.S.A.). Personal communication to the principal author (1970).

Liquat Ali, S.K., and Mazumdar, B.C. (1991). Studies on pectin content of some fruits and plant parts. *Science and Culture*. 57 (10 and 11): 256.

Lowry, O.H., Rosenbrough, N.J., Farr, A.L. and Randall, D.J. (1951). Protein measurement with the Folin-phenol reagent. *Journal of Biological Chemistry*. 193: 265-275.

Majumder, K. and Majumdar, B.C. (1996). Protein, fat and carbohydrate content in the mature seeds of jackfruit (*Artocarpus heterophyllus* Lamk.) *Applied Nutrition*. 21 (2): 22.

Majumder, K. and Mazumdar, B.C. (1998). Changing pattern of pectic substances in developing fruits of *Psidium guajava* L. *Indian Journal of Plant Physiology*. 3 (1): 42-45

Majumder, K. and Mazumdar, B.C. (2001). Effects of auxin and gibberellin on pectic substances and their degarding enzymes in developing fruits of cape-gooseberry (*Physalis peruviana* L.). *Journal of Horticultural Science & Biotechnology*. 76 (3): 276-279.

Malik, C.P. and Singh, M.B. (1980). *Plant Enzymology and Histoenzymology*. Kalyani Publishers, New Delhi, India.

Mammen, D. (1991). *Methods in Plant Chemistry and Economic Botany*. Kalyani Publishers, New Delhi.

Mandal, U. and Mazumdar, B.C. (1998). Qualitative constituents in some tropical and sub-tropical minor fruits during development. *Indian Biologist*. 30 (1): 70-80.

Mazumdar, B.C. (1975). Free- amino acids level in leaves of mango var. Dusheri. *Plant Science*. 7: 91.

Mazumdar, B.C. (1976). Physico-chemical changes in fruits of Kinnow and Nagpur Santra during development and maturation. *The Punjab Horticultural Journal*. 16 (3 and 4): 96-100.

Mazumdar, B.C. (1986). Qualitative behaviour of different types of emblica (*Phyllanthus emblica* L.) fruits and loss of their vitamin C by steeping with brine water. *Indian Biologist*. 18 (1): 34- 36.

Mazumdar, B.C. and Ali, Sk.L. (1998). *Production Technology of Papaya Latex*. New Age International (P) Ltd. Publishers, New Delhi, India.

Mazumdar, B.C. (2000). Panchanan – A wonder variety of mango evolved as a chance seedling. *Science and Culture*. 66 (3 and 4): 165-166.

McCready, R.M. and McComb, E.A. (1952). Extraction and determination of total pectic materials in fruits. *Analytical Chemistry*. 24: 1986-1988.

Mertz, E. T. (1951). *Elementary Biochemistry*. Vakils, Feffer & Simmons Pvt. Ltd., Bombay, India.

Monselise, S.P. (1986). *CRC Handbook of Fruit Set and Development*. CRC Press, Inc., Boca Raton, Florida, U.S.A.

Mukhopadhya, Paramita and Mazumdar, B.C. (1999). Nutritive value of seasonal fruits sold in Calcutta markets. *Science and Culture.*, 65 (3 and 4): 159-162 .

Murata, T., Akazawa, T. and Fukuchi, S. (1968). Enzymic method of starch breakdown in germinating rice seeds. An analytical study. *Plant Physiology*. 43: 899-905.

Nelson, N. (1944). A photometric adaptation of the Somogyi method for the determination of glucose. *Journal of Biological Chemistry*, 153: 373-380.

Northcote, D.H. (1972). Chemistry of the plant cell-wall. *Annual Review of Plant Physiology*. 23: 113-132.

Pal, D.K. (1995). (Plant Physiologist, Indian Institute of Horticultural Research, Bangalore, India). Personal communication to the authors.

Plummer, D.T. (1979). *An Introduction to Practical Biochemistry*. Tata McGraw-Hill Publishing Company Ltd., New Delhi, India.

Pressey, R. (1969). Potato sucrose synthetase: purification, properties and changes in activity associated with maturation. *Plant Physiology*. 44: 759.

Pucher, G.W., Leavenworth, G.S. and Vickery, H.B. (1948). Determination of starches in plant tissue. *Analytical Chemistry*. 20: 850-853.

Ranganna, S. (1986). *Hand Book of Analysis and Quality Control for Fruit and Vegetable Products*. Tata Mc-Graw Hill Publishing Company Ltd., New Delhi, India.

Ranganna, S. (Retired Scientist, Central Food Technological Research Institute, Mysore, India). Personal communication to the authors (1995).

Redgwell, R.J., Melton, L.D. and Brasch, D.J. (1992). Cell wall dissolution in ripening kiwifruit (*Actinidia deliciosa*) solublization of the pectic polymer. *Plant Physiology*. 98 (1): 71-81.

Roe, J.H. (1934). A colorimetric method for the determination of fructose in blood and urine. *Journal of Biological Chemistry*. 107: 15.

Ruck, J.A. (1963). *Ascorbic acid*. Publication 1154, Canada Department of Agriculture.

Sadasivam, S. and Manickam, A. (1992). *Biochemical Methods for Agricultural Science*. Wiley Eastern Limited, New Delhi, India.

Singh, S., Krishnamurthi, S. and Katyal, S.L. (1967). *Fruit Culture in India*. Indian Council of Agricultural Research, New Delhi, India.

Somogyi, M. (1952). Notes on sugar determinations. *Journal of Biological Chemistry*. 195: 19-23.

Stroev, E.A. (1986). *Biochemistry*. Mip Publishers, Moscow, U.S.S.R.

Tripathi, G. (1999). *Enzyme Biotechnology*. Technoscience Publishers New Delhi, India.

Wilson, K., Goulding, K.H. (1992). *A Biologist's Guide to Principle and Techniques of Practical Biochemistry*. Cambridge University Press, U.K.

Yemn, E.W. and Wills, A.J. (1954). The estimation of carbohydrates in plant extracts by anthrone. *Biochemical Journal*. 57: 508-514.

Yoshida, S., Porno, D.A., Cock, J.H. and Gomez, K.A. (1972). *Laboratory Manual For Physiological Studies of Rice*. The International Rice Research Institute, Los Banos, Laguna, Phillippines.

SUBJECT INDEX